Universitext

Universitext

Universitext is a series of textbooks that presents material from a wide variety of mathematical disciplines at master's level and beyond. The books, often well class-tested by their author, may have an informal, personal, even experimental approach to their subject matter. Some of the most successful and established books in the series have evolved through several editions, always following the evolution of teaching curricula, into very polished texts.

Thus as research topics trickle down into graduate-level teaching, first textbooks written for new, cutting-edge courses may make their way into *Universitext*.

More information about this series at http://www.springer.com/series/223

Matej Brešar

Introduction
to Noncommutative
Algebra

 Springer

Matej Brešar
University of Ljubljana
Ljubljana
Slovenia

and

University of Maribor
Maribor
Slovenia

ISSN 0172-5939   ISSN 2191-6675   (electronic)
ISBN 978-3-319-08692-7   ISBN 978-3-319-08693-4   (eBook)
DOI 10.1007/978-3-319-08693-4

Library of Congress Control Number: 2014943746

Mathematics Subject Classification: 16-01, 16Dxx, 16Kxx, 16Nxx, 16Pxx, 16Rxx, 16Sxx, 16Uxx

Springer Cham Heidelberg New York Dordrecht London

Printed on acid-free paper

Springer is part of Springer Science+Business Media (www.springer.com)

In memory of my father

Preface

The purpose of this book is to give a gentle introduction to noncommutative rings and algebras that requires fewer prerequisites than most other books on the subject. It is based on a series of lectures given to masters students at the University of Ljubljana and is intended to serve as a first reading on noncommutative algebra for beginning graduate and advanced undergraduate students. The first two chapters, in which some results of historic importance are derived by rudimentary tools, can have a wider audience. For example, they can be used in a course for future high school teachers, or for an undergraduate seminar. Mathematicians working in areas that only have some interactions with noncommutative algebra may also benefit from this book in which important, classical themes are presented in a simple manner, avoiding excessive generality.

The necessary background needed to follow this text is a standard knowledge of linear algebra and a basic knowledge about groups, rings, and fields. To make precise what we mean by this, the book begins with a survey of prerequisites. This is followed by the first chapter which considers finite dimensional division algebras. Its most prominent results are Frobenius' theorem on real division algebras and Wedderburn's theorem on finite division rings. The second chapter is devoted to the structure of finite dimensional algebras, featuring the classical Wedderburn's theory. After the first two warm-up chapters, which mostly deal with results known for a hundred years, the next two introduce and study more abstract notions: modules, vector spaces over division rings, and tensor products. The Double Centralizer Theorem and the Skolem-Noether Theorem are included therein. The fifth chapter considers the structure theory of rings. The main themes are primitive rings, the Jacobson Density Theorem, and the Jacobson radical. If the first five chapters survey the "greatest hits" of noncommutative algebra, the last two are slightly more specialized, reflecting my personal taste and interests. They treat polynomial identities and related notions, such as free algebras and rings of quotients.

The order of the topics in the book is quite close to the chronological order of their development. This actually occurred unintentionally, probably as a result of my attempt to follow the principle that an advanced concept should be introduced only when truly needed, and not at a higher level of generality than necessary. Not because I do not appreciate abstract concepts; after all, they are what makes mathematics beautiful. But it is because they can be better understood and valued when given some evidence that they are indispensable. Another principle I have followed, especially in the early chapters, is to choose proofs that appear to be the simplest. This may not always be the same as the shortest, but rather proofs that seem very natural and easy to memorize. It has been my desire to write the book in such a way that readers would not get an impression that one needs supernatural abilities to create mathematics, but that even they themselves may be able, with some luck and courage, to discover a little piece of it.

One challenge in the exposition of the book was to find new proofs of classical theorems that would fulfill the simplicity criteria described above. This has been in fact my little passion over the last years, which has eventually yielded some results. The proofs of the following theorems are different from those in the standard sources: Frobenius' theorem (Sect. 1.1), the Skolem-Noether Theorem (Sect. 1.6), Wedderburn's structure theorems (Sect. 2.9), Martindale's theorem on prime GPI-rings (Sect. 7.7), Posner's theorem on prime PI-rings (Sect. 7.9), and the Formanek-Razmyslov Theorem on central polynomials (Sect. 7.10). How much originality is there in these proofs? Frankly, I do not know. I have not found such proofs in the literature, and have published them in mathematical journals through a series of papers. On the other hand, so many mathematicians have known these theorems, especially the first three listed, for so many years that one hardly imagines that something entirely new about them can still be found. It is not so rare that some mathematical ideas are discovered, forgotten, and rediscovered. Anyway, I hope that these proofs provide interesting alternatives to those from other sources. Two remarks must be added. First, the proof of Frobenius' theorem was obtained in a joint work with Peter Šemrl and Špela Špenko. Second, shortly after the publication of my paper on alternative proofs of Wedderburn's structure theorems Edmund Puczylowski informed me that he had used them in his class, but with a modification—which I immediately liked. I am now happy to use this modified version in the book.

Each chapter ends with exercises of varying difficulty levels. They are sorted by topics, not by difficulty. Some of the harder ones are accompanied with hints, but others are not, in order to give the student the opportunity to fully enjoy the pleasure of discovering the solution. A fair amount of exercises are original, but there are also some that appear, either as exercises or as theorems with proofs, in almost every book on the subject. I tried to avoid long, theoretical exercises, and gave preference to those that I had found entertaining and appealing.

The list of references at the end of the book is very short. It includes only books which are explicitly referred to on relatively rare occasions when details are not provided (say, when closing the discussion on a topic with additional information, or when omitting tedious details in some example). Most of these books are textbooks on noncommutative rings and algebras, but written at a more advanced level than this one. An exception is perhaps J. Beachy's *Introductory Lectures on Rings and Modules* [Bea99] whose emphasis, however, differs from ours. At the time of writing I was frequently consulting many of the listed books, but mainly T. Y. Lam's *A First Course in Noncommutative Rings* [Lam01] (together with the accompanying problem book [Lam95]) and L. H. Rowen's *Graduate Algebra: Noncommutative View* [Row08]. I warmly recommend these two texts as further reading to students who want to dig deeper into the field of noncommutative algebra.

English is not my mother tongue and I apologize to the reader if this is sometimes too obvious. "Translating" the English stuck in my mind into the English as it should be has been a big and often frustrating challenge for me during the writing process. The bright side of this is that it gives me the opportunity to thank my sons Jure and Martin for their continuous help in this matter.

The most difficult step in writing a book is forcing yourself to write the first sentence. I am thankful to my friend and colleague Peter Šemrl for giving me the necessary encouragement to make it.

Rough drafts of most chapters were first read by Špela Špenko. Her criticism has been very helpful (as always). Aljaž Zalar has solved most of the exercises and made me think of modifying some. Igor Klep provided many helpful suggestions concerning Chaps. 1 and 6. The book has been read cover to cover by Nik Stopar and Janez Šter. Their comments have contributed to the improvement of the text and saved me from some errors. My sincere thanks go to all of them, as well as to the Springer staff for their professional assistance.

Ljubljana and Maribor, March 2014   Matej Brešar

Contents

Symbols

$\mathbb{N}$	Set of positive integers
$\mathbb{Z}$	Ring of integers
$\mathbb{Z}_n$	Ring of integers modulo n
$\mathbb{Q}$	Field of rational numbers
$\mathbb{R}$	Field of real numbers
$\mathbb{C}$	Field of complex numbers
$\mathbb{H}$	Division algebra of quaternions
δ_{ij}	Kronecker delta
$|S|$	Cardinality of the set S
$S \setminus T$	Set difference
id_S	Identity map on S
$\det(A)$	Determinant of the matrix A
$\mathrm{tr}(A)$	Trace of the matrix A
M^*	Group of invertible elements in the monoid M
S_n	Symmetric group on $\{1, \ldots, n\}$
$\mathrm{sgn}(\sigma)$	Sign of the permutation σ
$R \cong R'$	Isomorphic rings (or algebras, modules, etc.)
$\ker \varphi$	Kernel of the homomorphism φ
$\mathrm{im}\, \varphi$	Image of the homomorphism φ
R/I	Factor ring
IJ	Product of ideals
$[a, b]$	Commutator of a and b
$Z(R)$	Center of the ring R
$\mathrm{char}(R)$	Characteristic of R
R°	Opposite ring of R
$M_n(R)$	Ring of all $n \times n$ matrices over R
E_{ij}	Standard matrix units
e_{ij}	Matrix units

$\Pi_{i \in I} R_i$	Direct product of a family of rings
$R_1 \times \cdots \times R_n$	Direct product of a finite family of rings
$R[\omega]$	Ring of polynomials over R in one indeterminate
$R[\Omega]$	Ring of polynomials over R in indeterminates from the set Ω
$R[[\omega]]$	Formal power series ring over R
$\mathrm{soc}(R)$	Socle of R
$\mathrm{rad}(R)$	Jacobson radical of R
$\dim_F V$	Dimension of the vector space V over the field F
$[A : F]$	Dimension of the algebra A over F
$T_n(F)$	Algebra of upper triangular $n \times n$ matrices over F
$C[a, b]$	Algebra of continuous functions from $[a, b]$ to $\mathbb{R}$
$\mathscr{A}_n$	nth Weyl algebra
$F[G]$	Group algebra of the group G over F
L_a	Left multiplication map $x \mapsto ax$
R_b	Right multiplication map $x \mapsto xb$
$M(A)$	Multiplication algebra of the algebra A
$A^\sharp$	Unitization of A
A_K	Scalar extension of A to the field K
$C_A(S)$	Centralizer of the set S in A
$\mathrm{ann}_R(M)$	Annihilator of the R-module M
$\mathrm{End}_R(M)$	Ring of endomorphisms of the R-module M
$\mathrm{End}_D(V)$	Ring of linear operators of the vector space V over the division ring D
$\mathrm{End}_F(V)$	Algebra of linear operators of the vector space V over the field F
$\oplus_{i \in I} M_i$	Direct sum of a family of submodules (or additive subgroups, or subspaces)
$M_1 \oplus \cdots \oplus M_n$	Direct sum of a finite family of submodules (or additive subgroups, or subspaces)
$M_1 + \cdots + M_n$	Sum of a finite family of submodules (or additive subgroups, or subspaces)
$U \otimes V$	Tensor product of vector spaces
$u \otimes v$	Simple tensor
$\varphi \otimes \psi$	Tensor product of linear maps
$A \otimes B$	Tensor product of algebras
$F\langle \xi, \eta \rangle$	Free algebra in two indeterminates
$F\langle X \rangle$	Free algebra in indeterminates from the set X
$\deg(f)$	Degree of the polynomial f
s_n	Standard polynomial
c_n	Capelli polynomial
$\mathrm{Id}(A)$	T-ideal of all polynomial identities of the algebra A
$GM_n(F)$	Algebra of generic matrices
$\widehat{Z}$	Field of quotients of the commutative domain Z

$F(\Omega)$	Field of rational functions in Ω
$Q_Z(R)$	Ring of central quotients of the ring R
$Q_{rc}(R)$	Right classical ring of quotients of R
$Q_r(R)$	Right Martindale ring of quotients of R
R_C	Central closure of R

Prerequisites

This preliminary chapter aims to give an accurate survey of the background material needed to follow this text. It is supposed to be used as a handbook, only when there is a need. We will therefore write it in a concise and condensed manner, paying less attention to motivation and insight than in the rest of the book. Some simpler proofs that are as much a part of understanding the subject as definitions and theorems will be outlined, but deeper results will be just stated as facts. The reader should be warned that the topics will be presented in a slightly unbalanced manner. Some will be examined carefully, while other equally or even more important ones will be only touched upon or even omitted. Of course, the emphasis will be on topics that are needed later.

Through this chapter we also indicate what knowledge the reader is assumed to have. We gain different knowledge at different places, so there might be at least something new for many of the readers. Maybe we will also mention a couple of things for which it would be pretentious to say that everyone should have learnt them in one of the undergraduate courses. Some notational and terminological conventions will also be included. Therefore, even well-prepared readers may occasionally consult this chapter, say in case they find something ambiguous in the text. Most of the conventions, however, are standard. Let us just point out two issues before we start.

Formally, the term "noncommutative ring" means a not *necessarily* commutative ring. But we shall use this term in a loose manner. By saying that this book is about noncommutative rings we mean that it considers rings that can be commutative or not, while by saying that some particular ring is noncommutative we shall mean that this ring is *not* commutative. At any rate, even when we use this term in its precise sense, the stress is on rings that are not commutative. The goals and the means of the theory of commutative rings are somewhat different.

We will not assume that every ring must have a unity 1, which may be a subject of dispute. Not that this is uncommon, but the fact is that many mathematicians believe that, at least in a textbook, the existence of 1 should be added to the axioms of a ring. This has advantages, but also disadvantages. There are some interesting and important classes of non-unital rings, e.g., nilpotent rings, nil rings, radical rings, and simple rings consisting of finite rank operators. In the non-unital setting we can treat ideals as (sub)rings and we can avoid the unitization process for rings that seem more natural without unity. On the other hand, sometimes the non-unital situation is not essentially more interesting, but causes technical difficulties which are distracting for a newcomer. At such instances we will simply add the assumption that a ring has a unity. Of course, some price has to be paid for omitting a possible axiom; admittedly, here and there our exposition would be slightly simpler if the presence of unity was automatically assumed. The readers who believe that all rings should have unities can simply skip parts of the text where the lack of unity creates some problems. There are not that many anyway.

Sets and Relations

First, a word about notation. By the set of natural numbers, $\mathbb{N}$, we mean positive (not nonnegative) integers:

$$\mathbb{N} = \{1, 2, \ldots\}.$$

We assume there is no need to explain the meaning of $\emptyset, \cap, \cup, \subseteq, \supseteq$. By $\subsetneq$ and $\supsetneq$ we denote the strict inclusion, by $S \setminus T$ the set $\{s \in S \mid s \notin T\}$, and by $|S|$ the cardinality of the set S.

If I is a (possibly infinite) index set and $\{S_i \mid i \in I\}$ is a family of sets, then $\Pi_{i \in I} S_i$ denotes the **Cartesian product** of the sets S_i. Thus, $\Pi_{i \in I} S_i$ is the set of all maps $f : I \to \bigcup_{i \in I} S_i$ such that $f(i) \in S_i$ for every i. Setting $s_i := f(i)$ we can write f as (s_i). If I is finite, say $I = \{1, \ldots, n\}$, then we write (s_i) as $(s_1, \ldots, s_n)$, and $\Pi_{i \in I} S_i$ as $S_1 \times \cdots \times S_n$. In particular, $S_1 \times S_2$ is the set $\{(s_1, s_2) \mid s_1 \in S_1, s_2 \in S_2\}$.

By id_S we denote the identity map on the set S. Given a map $f : X \to Y$ and sets $S \subseteq X$, $T \subseteq Y$, we write $f(S)$ for $\{f(s) \mid s \in S\}$ and $f^{-1}(T)$ for $\{x \in X \mid f(x) \in T\}$. By $f|_S$ we denote the restriction of f to S. If f is bijective, then f^{-1} denotes its inverse map. The function composition will be denoted by fg rather than $f \circ g$. The reason behind this is that functions appearing in algebra are often considered as elements of an algebraic system in which multiplication is defined as composition.

A relation $\sim$ on a set S is said to be an **equivalence relation** if it is reflexive, symmetric, and transitive. This means that $a \sim a$, if $a \sim b$ then $b \sim a$, and if $a \sim b$ and $b \sim c$ then $a \sim c$. The equivalence class of an element a is the set of all $b \in S$ such that $a \sim b$. It is easy to see that S is a disjoint union of its equivalence classes. An equivalence relation on S thus gives rise to a partition of S.

Our next goal is to state a somewhat less elementary, yet one of the most frequently used results in mathematics—the so-called Zorn's lemma. A set S together with a relation $\leq$ which is reflexive ($a \leq a$), anti-symmetric ($a \leq b$ and $b \leq a$ implies $a = b$), and transitive ($a \leq b$ and $b \leq c$ implies $a \leq c$) is called a **partially ordered set**. For example, the power set of a given set is partially ordered by inclusion. A subset C of a partially ordered set S is called a **chain** if for each pair $a, b \in C$ we have either $a \leq b$ or $b \leq a$. An element $u \in S$ is called an **upper bound** for C if $c \leq u$ for every $c \in C$. Finally, an element $m \in S$ is called a **maximal element** of S if for every $a \in S$, $m \leq a$ implies $m = a$. We are now in a position to state

Zorn's lemma *Let S be a nonempty partially ordered set. If every chain in S has an upper bound in S, then S contains a maximal element.*

Zorn's lemma is equivalent to the **axiom of choice** which says the following: If $\{A_i \mid i \in I\}$ is a family of nonempty sets, then there exists a family of elements $\{a_i \mid i \in I\}$ such that $a_i \in A_i$ for every $i \in I$.

Groups

A **semigroup** is a set S together with an associative binary operation. This means that S is endowed with a map $S \times S \to S$, $(x, y) \mapsto xy$, such that $(xy)z = x(yz)$ for all $x, y, z \in S$. It is easy to see that associativity makes it possible for us to write all products, including those of more than three elements, without parentheses. In particular, we can define the nth power x^n of every $x \in S$ and every $n \in \mathbb{N}$. We say that elements $x, y \in S$ commute if $xy = yx$. If $xy = yx$ holds for all $x, y \in S$, then we say that S is commutative.

An element $e \in S$ that satisfies $ex = x$ for every $x \in S$ is called a left identity element. Similarly, a right identity element is every element $e' \in S$ satisfying $xe' = x$ for every $x \in S$. If S has both a left identity element e and a right identity element e', then they are equal; indeed, they both must be equal to ee'. In this case we call $e = e'$ an **identity element** of S. From now on we will denote an identity element by 1.

A semigroup with a (necessarily unique) identity element 1 is called a **monoid**. An element a in a monoid M is said to be **right invertible** if there exists $b \in M$ such that $ab = 1$. Every such element b is called a **right inverse** of a. Analogously we define left invertibility and left inverses. We say that a is **invertible** if it is both right and left invertible. In this case any right inverse of a is equal to any left inverse of a. Indeed, $ab = ca = 1$ implies $b = (ca)b = c(ab) = c$. Thus, there exists a unique element which is both a left and right inverse of a. We call it the **inverse** of a and denote it by a^{-1}. If a is invertible, then so is a^n for every $n \in \mathbb{N}$. Its inverse is $(a^{-1})^n$ which we also denote by a^{-n}. For convenience we set $a^0 = 1$ for every $a \in M$. Further, if a and b are invertible, then so is ab, and $(ab)^{-1} = b^{-1}a^{-1}$. The set of all invertible elements of M will be denoted by M^*.

A monoid in which every element is invertible is called a **group**. In the next paragraphs we give some basic examples.

If M is a monoid, then M^* is clearly a group. For example, $\mathbb{Z}$ is a monoid under (the usual) multiplication, and hence $\mathbb{Z}^* = \{-1, 1\}$ is a group. Similarly, if $\mathbb{F} \in \{\mathbb{Q}, \mathbb{R}, \mathbb{C}\}$ then $\mathbb{F}$ is a monoid under multiplication, and $\mathbb{F}^* = \mathbb{F} \setminus \{0\}$. The set $M_n(\mathbb{F})$ of all $n \times n$ matrices with entries in $\mathbb{F}$ is a monoid under the usual matrix multiplication; the corresponding group $M_n(\mathbb{F})^*$ of all invertible matrices is called the **general linear group** (over $\mathbb{F}$).

Let X be a set. A bijective map from X onto X is called a **permutation**. The set of all permutations of X is a group under the composition. Indeed, the composition is associative, id_X is the identity element, and inverses clearly exist. We call it the **symmetric group** of X. If $X = \{1, \ldots, n\}$, then we denote the symmetric group of X by S_n. It is easy to see that $|S_n| = n!$. A permutation in S_n is called a transposition if it interchanges two of the elements and leaves all the rest fixed. Every permutation $\sigma \in S_n$ can be written as a product of transpositions. If $\sigma = \tau_1 \ldots \tau_p = \rho_1 \ldots \rho_q$, where all τ_i and ρ_i are transpositions, then either both p and q are even or both p and q are odd. In the first case we say that σ is an even permutation, and in the second case we say that it is an odd permutation. We define the **sign** of σ, $\mathrm{sgn}(\sigma)$, by $\mathrm{sgn}(\sigma) := \begin{cases} 1 & \text{if } \sigma \text{ is even} \\ -1 & \text{if } \sigma \text{ is odd} \end{cases}$.

A commutative group is called **abelian**. So far we have used only multiplicative notation for binary operations. For abelian groups we often use additive notation. This means that the result of the operation is written as $x + y$ instead of xy. An abelian group in which this notation is used is called an **additive group**. (The convention that an additive group must necessarily be abelian is not entirely standard, but it is convenient to adopt it in this book.) An identity element in an additive group is written as 0. Further, we write $-x$ instead of x^{-1} and nx instead of x^n. A simple example of an additive group is $\mathbb{Z}$ under the usual addition. Another example is $\mathbb{Z}_n = \{0, 1, \ldots, n - 1\}$ where addition is taken modulo n.

Until further notice we continue using the multiplicative notation. A map φ from a group G into a group G' is called a **group homomorphism**, or simply a **homomorphism** when it is clear from the context that groups are the subject of consideration, if $\varphi(xy) = \varphi(x)\varphi(y)$ for all $x, y \in G$. The set $\ker\varphi := \{x \in G \mid \varphi(x) = 1\}$ is called the **kernel** of φ, and the set $\mathrm{im}\,\varphi := \{\varphi(x) \mid x \in G\} \, (= \varphi(G))$ is called the **image** of φ. Note that φ is injective if and only if $\ker\varphi = \{1\}$.

A bijective homomorphism is called an **isomorphism**. We say that groups G and G' are isomorphic, and write $G \cong G'$, if there exists an isomorphism φ from G onto G'. In this case its (set-theoretic) inverse φ^{-1} is an isomorphism from G' onto G. The relation of being isomorphic is an equivalence relation. Isomorphic groups have exactly the same properties; informally we consider them as identical, although they may appear very different at a glance. For example, consider the set G of all matrices of the form $\begin{bmatrix} \alpha & \beta \\ -\beta & \alpha \end{bmatrix}$ with $\alpha, \beta \in \mathbb{R}$, $\alpha \neq 0$ or $\beta \neq 0$. One can check that G is a group under multiplication, and that the map

$$\begin{bmatrix} \alpha & \beta \\ -\beta & \alpha \end{bmatrix} \mapsto \alpha + \beta i$$

is an isomorphism from G onto $\mathbb{C}^* = \mathbb{C} \setminus \{0\}$. Thus, as long as we are interested only in multiplication, the set of matrices G is just a disguised form of the more familiar set of nonzero complex numbers. Another example: The multiplicative group $(0, \infty)$ is isomorphic to the additive group $\mathbb{R}$ via the isomorphism $x \mapsto \ln(x)$.

A homomorphism from a group into itself is called an **endomorphism**. An **automorphism** is an endomorphism which is also an isomorphism. An injective homomorphism $\varphi : G \to G'$ is sometimes called an **embedding**. This term is usually used when we wish to point out that by identifying G with im φ we may view G as a subgroup of G'. We also say that G embeds in G'. For example, the additive group $\mathbb{R}$ can be canonically embedded into the additive group $\mathbb{C}$ by $\alpha \mapsto \alpha + 0i$.

Every group G can be embedded into a symmetric group. This result is known as **Cayley's theorem**, and can be proved as follows. For any $a \in G$ we define $\ell_a : G \to G$ by $\ell_a(g) = ag$. One immediately checks that ℓ_a is a permutation of G, $\ell_a = \mathrm{id}_G$ if and only if $a = 1$, and $\ell_{ab} = \ell_a \ell_b$ for all $a, b \in G$. Accordingly, $a \mapsto \ell_a$ is an embedding of the group G into the symmetric group of the set G.

A homomorphism between additive groups is also called an **additive map**. We will also come across biadditive maps: We say that a map $f : G_1 \times G_2 \to G'$, where G_1, G_2, G' are additive groups, is **biadditive** if it is additive in each variable, i.e.,

$$f(x + y, z) = f(x, z) + f(y, z) \quad \text{and} \quad f(x, z + w) = f(x, z) + f(x, w)$$

for all $x, y \in G_1$, $z, w \in G_2$.

A subset H of a group G is called a **subgroup** if it contains 1, is closed under the group operation, i.e., $h, h' \in H$ implies $hh' \in H$, and $h^{-1} \in H$ whenever $h \in H$. Note that H is itself a group. A (left) **coset** of H in G is a set $xH := \{xh \mid h \in H\}$ where x is a fixed element in G. It is easy to see that $|xH| = |yH|$ for all $x, y \in G$, $xH = yH$ if and only if $x^{-1}y \in H$, and $xH \neq yH$ if and only if $xH \cap yH = \emptyset$. From these observations one easily infers **Lagrange's theorem** which says that the number of cosets of a subgroup H in a finite group G is equal to $\frac{|G|}{|H|}$. In particular, $|H|$ thus divides $|G|$. Let us point out one corollary to this fundamental result. We say that an element a in a group G has **finite order** if $a^r = 1$ for some $r \in \mathbb{N}$. If r is the smallest natural number with this property, then r is called the **order** of a. In this case $\langle a \rangle := \{1, a, \ldots, a^{r-1}\}$ is a subgroup of G, called the cyclic subgroup generated by a. If G is finite, then every element in G is of finite order, and hence Lagrange's theorem tells us that $|\langle a \rangle|$ divides $|G|$ for every $a \in G$. Since $|\langle a \rangle| = r$ is the order of a it follows that $a^{|G|} = 1$ for every $a \in G$.

Let a be an arbitrary element in a group G. Every element of the form xax^{-1}, $x \in G$, is called a **conjugate** of a. A subgroup N of G is called a **normal** subgroup if it contains every conjugate of each of its elements, i.e., $a \in N$ implies $xax^{-1} \in N$ for every $x \in G$. Every subgroup of an abelian group is normal. The **center** of a group G, defined as the set of all elements in G that commute with every other element in G, is a normal subgroup of G. If $\varphi : G \to G'$ is a homomorphism, then

$\ker \varphi$ is a normal subgroup of G, while $\operatorname{im} \varphi$ is a subgroup of G' which is not necessarily normal.

The next construction works for normal subgroups of arbitrary groups, but we confine ourselves to additive groups as this is all we need. Thus, let H be a subgroup of an additive group G. A coset of H in G is in the additive notation written as $x + H$. Let G/H denote the set of all cosets of H in G endowed with the addition defined by

$$(x + H) + (y + H) := (x + y) + H.$$

It is easy to see that this operation is well-defined, i.e., $x + H = x' + H$ and $y + H = y' + H$ implies $(x + y) + H = (x' + y') + H$, and that G/H is a group under this operation. We call it the **factor group** of G by H. The map $x \mapsto x + H$ is a surjective homomorphism from G onto G/H with the kernel H. It is called the **canonical homomorphism**.

Let G' be another additive group. If $\varphi : G \to G'$ is a homomorphism, then

$$G/\ker \varphi \cong \operatorname{im} \varphi.$$

An isomorphism is given by $x + \ker \varphi \mapsto \varphi(x)$. Indeed, this map is well-defined since $x + \ker \varphi = y + \ker \varphi$ implies $x - y \in \ker \varphi$ and hence $\varphi(x) = \varphi(y)$. Clearly, it is additive and a bijection from $G/\ker \varphi$ onto $\operatorname{im} \varphi$.

Given groups $G_1, \ldots, G_n$, we define their **direct product** $G_1 \times \cdots \times G_n$ as the Cartesian product of the sets $G_1, \ldots, G_n$ endowed with the componentwise operation:

$$(x_1, \ldots, x_n)(y_1, \ldots, y_n) = (x_1 y_1, \ldots, x_n y_n).$$

It is immediate that $G_1 \times \cdots \times G_n$ is a group, and that each G_i can be naturally (canonically) embedded in $G_1 \times \cdots \times G_n$. The direct product $\Pi_{i \in I} G_i$ of a possibly infinite family of groups $\{G_i \mid i \in I\}$ is defined analogously. The operation is thus given by $(x_i)(y_i) = (x_i y_i)$. If G_i are additive groups, then the operation in $\Pi_{i \in I} G_i$ is also written additively: $(x_i) + (y_i) = (x_i + y_i)$.

Let now $G_1, \ldots, G_n$ be subgroups of an additive group G. Then so is their sum

$$G_1 + \cdots + G_n := \{x_1 + \cdots + x_n \mid x_1 \in G_1, \ldots, x_n \in G_n\}.$$

If

$$G_i \cap (G_1 + \cdots + G_{i-1} + G_{i+1} + \cdots + G_n) = \{0\}$$

for every i, then $G_1 + \cdots + G_n$ is called the **direct sum** of the subgroups G_i, and is denoted by $G_1 \oplus \cdots \oplus G_n$. Every element in $G_1 \oplus \cdots \oplus G_n$ can be written as $x_1 + \cdots + x_n$ where $x_i \in G_i$ in a unique way, so the x_i's basically play the role of components. The difference between the groups $G_1 \oplus \cdots \oplus G_n$ and $G_1 \times \cdots \times G_n$ is merely formal. In more accurate terms, they are canonically isomorphic via $x_1 + \cdots + x_n \mapsto (x_1, \ldots, x_n)$.

Rings

A **ring** R is an additive group endowed with another associative binary operation $(x, y) \mapsto xy$, called multiplication, such that the distributive laws hold in R, i.e.,

$$(x + y)z = xz + yz \quad \text{and} \quad x(y + z) = xy + xz$$

for all $x, y, z \in R$. In other words, $(x, y) \mapsto xy$ is a biadditive map. The following simple facts follow easily from the definition: $0x = x0 = 0$, $(-x)y = x(-y) = -xy$, $(-x)(-y) = xy$ for all $x, y \in R$.

A ring R is **commutative** if $xy = yx$ for all $x, y \in R$. For example, $\mathbb{Z}$, with the usual addition and multiplication, is a commutative ring. The additive group $\mathbb{Z}_n$ becomes a commutative ring if we take multiplication modulo n. The set $C[a, b]$ of continuous functions $f : [a, b] \to \mathbb{R}$, endowed with the usual addition and multiplication of functions, is a commutative ring.

Commutative rings play a secondary role in this book. But we will be often concerned with the "most commutative" part of our rings. The set

$$Z(R) := \{c \in R \mid cx = xc \text{ for all } x \in R\}$$

is called the **center** of the ring R, and its elements are called **central elements**. Obviously, R is commutative if and only if $Z(R) = R$. If R is not commutative then $Z(R)$ is more often than not a relatively small subset of R.

We assume the reader is familiar with at least one example of a noncommutative ring, namely with $M_n(\mathbb{R})$, the ring of all $n \times n$ matrices with entries in $\mathbb{R}$. Here one takes the usual matrix addition and multiplication. We will introduce matrix rings in a more accurate and general manner in the next section.

One of the most intriguing properties in rings is that the product of two nonzero elements can be 0. This cannot happen in, say, $\mathbb{Z}$, but can easily occur in $\mathbb{Z}_n$ if n is not prime, or in $M_n(\mathbb{R})$. A nonzero element x in a ring R is said to be a **left zero-divisor** if there exists a nonzero element $y \in R$ such that $xy = 0$. Analogously we define a **right zero-divisor**. An element that is both a left and right zero-divisor is called a **zero-divisor**. For example, a central element that is a left (or right) zero-divisor is automatically a zero-divisor.

Every ring is a semigroup under multiplication. If it is actually a monoid, then it will be called a **unital ring**. Thus, a ring R is unital if it contains an element 1 such that $1x = x1 = x$ for all $x \in R$. We will call 1 the **unity** of R. Most rings that first come into our minds are unital; a simple example of a ring without unity is $2\mathbb{Z}$, the ring of even integers. If $1 = 0$ in a ring R, then R is the **trivial** or **zero ring** whose only element is 0.

In a unital ring we can speak about left invertible, right invertible, and invertible elements. A nonzero unital ring D in which every nonzero element is invertible (i.e., $D^* = D \setminus \{0\}$) is called a **division ring**. Division rings obviously have no (left, right) zero-divisors. A commutative division ring is called a **field**.

Obvious examples are $\mathbb{Q}$, $\mathbb{R}$, and $\mathbb{C}$. It is easy to show that $\mathbb{Z}_p$ with p prime is a field. Conversely, $\mathbb{Z}_n$ is not a field if n is not prime.

We say that a ring R has **characteristic** n if n is the smallest natural number such that $nx = 0$ for every $x \in R$ (note that in a unital ring R the condition $nx = 0$ for every $x \in R$ is equivalent to $n1 = 0$). If no such natural number n exists, then we say that R has **characteristic** 0. The characteristic of R will be denoted by char(R). For example, char$(\mathbb{Z}) = 0$ and char$(\mathbb{Z}_n) = n$. If R is unital and $m \in \mathbb{N}$ is such that $m1 = 0$, then $m = rs$ implies $(r1)(s1) = 0$. Hence, either R has zero-divisors or char(R) is prime. In particular, the characteristic of a field is either 0 or a prime number p. Fields of prime characteristic are called fields of **finite characteristic**. The subset F_0 of a field F consisting of all elements of the form $(n1)(m1)^{-1}$, $n, m \in \mathbb{Z}$, $m1 \neq 0$, is easily seen to be a field isomorphic to $\mathbb{Q}$ if char$(F) = 0$, and to $\mathbb{Z}_p$ if char$(F) = p$. We call F_0 the **prime subfield** of F.

A map φ from a ring R into a ring R' is called a **ring homomorphism** if

$$\varphi(x + y) = \varphi(x) + \varphi(y) \quad \text{and} \quad \varphi(xy) = \varphi(x)\varphi(y)$$

for all $x, y \in R$. Thus, φ is an additive group homomorphism and simultaneously a multiplicative semigroup homomorphism. Its **kernel** is $\ker \varphi := \{x \in R \mid \varphi(x) = 0\}$, and its **image** is $\operatorname{im} \varphi := \{\varphi(x) \mid x \in R\}$. The other necessary definitions are the same as for group homomorphisms (as well as for homomorphisms of other algebraic structures). Thus, an **isomorphism** is a bijective homomorphism, an **endomorphism** is a homomorphism from a ring into itself, an **automorphism** is an endomorphism which is also an isomorphism, and an **embedding** is an injective homomorphism. The relation of being isomorphic is an equivalence relation. We write $R \cong R'$ if R and R' are isomorphic. Let us modify our earlier example on groups to give one illustration. The set R of all matrices of the form $\begin{bmatrix} \alpha & \beta \\ -\beta & \alpha \end{bmatrix}$ with $\alpha, \beta \in \mathbb{R}$, is easily seen to be a ring isomorphic to $\mathbb{C}$. An isomorphism is defined exactly as before, just that now there is no need to exclude the case where $\alpha = \beta = 0$. Mathematicians are inclined to "forget" about the difference between isomorphic objects, so some might even say that R *is* the field of complex numbers.

A subset of a ring R which is a subgroup with respect to addition is called an **additive subgroup** of R. Let S be an arbitrary nonempty subset of R. The set $\overline{S}$ of all elements of the form $m_1 s_1 + \cdots + m_n s_n$ where $m_i \in \mathbb{Z}$ and $s_i \in S$ is an additive subgroup of R which contains S. Moreover, $\overline{S}$ is the smallest additive subgroup containing S in the sense that it is contained in every other additive subgroup which contains S. We call $\overline{S}$ the **additive subgroup generated by** S. (For $S = \emptyset$ we define $\overline{S} = \{0\}$.)

An additive subgroup of R which is closed under multiplication is called a **subring** of R. A subring is itself a ring. For example, the center $Z(R)$ is a subring of R which is commutative as a ring. We remark that $Z(R)$ can contain only 0 even when R is a nonzero ring. If, however, R is unital, then $1 \in Z(R)$. A subring D of a unital ring R that contains 1 and is a division ring is called a **division subring** of R. If it is also commutative, then we call it a **subfield** of R.

The subring generated by the subset S of R is, of course, defined as the smallest subring of R containing S. Note that it is equal to the additive subgroup generated by the set of all possible (finite) products $s_1 \ldots s_r$ of elements $s_i \in S$. If the subring generated by S is R itself, then we call S a **generating set** for the ring R, and its elements are called **generators** of R. We also say that S **generates** R. A ring is said to be **finitely generated** if it contains a finite generating set.

The terms "generated by", "generating set", "generators", "finitely generated" are used, in the same sense as above, for other algebraic structures as well. We hope that their meaning can be understood without always giving a detailed explanation.

A subring I of R is called a **left ideal** of R if $xu \in I$ for all $x \in R$ and $u \in I$. Analogously, a subring I of R is called a **right ideal** of R if $ux \in I$ for all $x \in R$ and $u \in I$. If I is both a left and right ideal, then I is called an **ideal**; we also say a **two-sided ideal** when we wish to stress that it is not only left or right. Left and right ideals are also called **one-sided ideals**.

For example, the ideals of the ring $\mathbb{Z}$ are $n\mathbb{Z} := \{nk \mid k \in \mathbb{Z}\}$, $n \geq 0$. In commutative rings the adjectives "left" and "right" are, of course, superfluous. The set of all matrices of the form $\begin{bmatrix} \alpha & 0 \\ \beta & 0 \end{bmatrix}$, $\alpha, \beta \in \mathbb{R}$, is an example of a left ideal but not an ideal of the ring $M_2(\mathbb{R})$. By putting zeros in one of the rows instead of in a column we obtain a right ideal that is not an ideal.

Every ring R has at least two ideals: $\{0\}$ and R. From now on we will write $\{0\}$ as 0. The reader should thus be warned that the symbol 0 can mean two different things, either the zero element or the zero (or trivial) ideal. From the context it should always be clear what we have in mind.

A (left, right, two-sided) ideal I different from R is said to be **proper**. Note that in this case I does not contain any invertible element if R is unital.

If I and J are ideals of R, then so is their intersection $I \cap J$, their **sum** $I + J := \{u + v \mid u \in I, v \in J\}$, and their **product** IJ which is defined as the additive subgroup generated by the set $\{uv \mid u \in I, v \in J\}$. The same is true if we replace "ideal" by "left ideal" or "right ideal" (incidentally, the sum and the product of subrings may not be a subring). We remark that, in accordance with the definition of the product, I^n should be understood as the additive subgroup generated by the set $\{u_1 \ldots u_n \mid u_i \in I\}$. Let us also point that $IJ \subseteq I \cap J$ if I, J are two-sided ideals.

Let S be a subset of R and let $a, b \in R$. We write

$$aS := \{ax \mid x \in S\}, \quad Sa := \{xa \mid x \in S\}, \quad aSb := \{axb \mid x \in S\}.$$

Note that Ra is a left ideal and aR is a right ideal. By RaR we denote the additive subgroup generated by the set $\{xay \mid x, y \in R\}$, which is clearly an ideal. If R is unital, then Ra is the left ideal generated by a, aR is the right ideal generated by a, and RaR is the ideal generated by a. If R is not unital, then the left ideal generated by a is equal to $Ra + \mathbb{Z}a$ where $\mathbb{Z}a := \{ka \mid k \in \mathbb{Z}\}$, the right ideal generated by a is equal to $aR + \mathbb{Z}a$, and the ideal generated by a, which is sometimes denoted by (a), is equal to $RaR + Ra + aR + \mathbb{Z}a$.

We have thus described the (left, right, two-sided) ideals generated by one element. Replacing one element by an arbitrary subset S of R does not make the task of description much harder. For example, the **ideal generated by** S is the additive subgroup of R generated by the set of elements of the form xsy, xs, sy, s with $x, y \in R$ and $s \in S$.

The **sum of an arbitrary family of ideals** $\{I_i \mid i \in I\}$ is defined as the ideal generated by $\bigcup_{i \in I} I_i$. It is easy to see that it is actually equal to the additive subgroup generated by $\bigcup_{i \in I} I_i$. The same is true for left or right ideals.

Let I be an ideal of a ring R. As I is, in particular, an additive subgroup of R, we can form the factor additive group R/I. Recall that R/I consists of all cosets $x + I$, $x \in R$, which are added according to

$$(x + I) + (y + I) = (x + y) + I.$$

We can make R/I into a ring by defining the multiplication by

$$(x + I)(y + I) := xy + I.$$

It is quite obvious that the ring axioms are fulfilled. The main issue is to show that this multiplication is well-defined. However, using $RI \subseteq I$ and $IR \subseteq I$ this is also easy. We call R/I the **factor ring** of R by I (or the quotient ring, but we will avoid this term to prevent confusion with the notion of a ring of quotients). Note that the canonical homomorphism $x \mapsto x + I$, mentioned in the additive group setting, is now a ring homomorphism.

A simple example: the ring $\mathbb{Z}/n\mathbb{Z}$ is isomorphic to $\mathbb{Z}_n$.

Let $\varphi : R \to R'$ be a ring homomorphism. Note that $\ker \varphi$ is an ideal of R, while $\operatorname{im} \varphi$ is in general only a subring of R'. Just as in the group-theoretic setting we have

$$R/\ker \varphi \cong \operatorname{im} \varphi$$

with an isomorphism defined by $x + \ker \varphi \mapsto \varphi(x)$.

Given rings $R_1, \ldots, R_n$, we define their **direct product** $R_1 \times \cdots \times R_n$ as the Cartesian product endowed with the componentwise operations:

$$(x_1, \ldots, x_n) + (y_1, \ldots, y_n) = (x_1 + y_1, \ldots, x_n + y_n),$$
$$(x_1, \ldots, x_n)(y_1, \ldots, y_n) = (x_1 y_1, \ldots, x_n y_n).$$

Clearly, $R_1 \times \cdots \times R_n$ is a ring. We may view R_i as an ideal of $R_1 \times \cdots \times R_n$ via the canonical embedding $x_i \mapsto (0, \ldots, 0, x_i, 0, \ldots, 0)$.

One analogously defines the direct product $\prod_{i \in I} R_i$ of a possibly infinite family of rings $\{R_i \mid i \in I\}$. The operations are thus given as follows:

$$(x_i) + (y_i) = (x_i + y_i),$$
$$(x_i)(y_i) = (x_i y_i).$$

Incidentally, the subset of $\prod_{i \in I} R_i$ consisting of all (x_i) such that all but finitely many x_i are 0 is a subring of $\prod_{i \in I} R_i$, called the **direct sum** of the family $\{R_i \mid i \in I\}$. The notions "direct product" and "direct sum" thus coincide for finite families of rings. It is therefore common to use the notation $R_1 \oplus \cdots \oplus R_n$ instead of $R_1 \times \cdots \times R_n$. However, we will stick with the latter.

Matrices

We have already mentioned that the set of all real $n \times n$ matrices, $M_n(\mathbb{R})$, is a ring under the usual matrix operations. This ring is not commutative as long as $n \geq 2$, and has plenty of zero-divisors. One can replace the role of $\mathbb{R}$ in the definition of addition and multiplication not only by another field, but by any other ring. This does not affect the proof that the ring axioms are fulfilled.

Let us be precise. Take an arbitrary ring R and $n \in \mathbb{N}$. Given any $a_{ij} \in R$, $1 \leq i, j \leq n$, we denote by (a_{ij}) the $n \times n$ matrix

$$\begin{bmatrix} a_{11} & a_{12} & \cdots & a_{1n} \\ a_{21} & a_{22} & \cdots & a_{2n} \\ \vdots & \vdots & \ddots & \vdots \\ a_{n1} & a_{n2} & \cdots & a_{nn} \end{bmatrix}.$$

Define addition and multiplication of such matrices by

$$(a_{ij}) + (b_{ij}) := (a_{ij} + b_{ij})$$

and

$$(a_{ij})(b_{ij}) := (c_{ij}), \text{ where } c_{ij} = \sum_{k=1}^{n} a_{ik} b_{kj}.$$

One readily checks that in this way the set of all $n \times n$ matrices with entries in R becomes a ring. It will be denoted by $M_n(R)$. Note that $M_1(R)$ can be identified with R. If R is unital, then so is $M_n(R)$. Its unity is the identity matrix I_n, i.e., the matrix whose entries on the main diagonal are 1 and all other entries are 0.

Sometimes we call $M_n(R)$ the **full matrix ring** over R, in order to point out that it contains all $n \times n$ matrices with entries in R. The term **matrix ring** is often used as a synonym for the full matrix ring, but it can also mean any ring whose elements are matrices.

A special but extremely important case is when $R = F$ is a field. When working with $M_n(F)$ we often rely on linear algebra methods. We assume the reader is familiar with the notions such as the eigenvalue, determinant, trace, and characteristic polynomial of a matrix.

Polynomials

Let R be an arbitrary ring R. We define a **polynomial** over R as a formal infinite sum

$$\sum_{i=0}^{\infty} a_i \omega^i = a_0 + a_1 \omega + a_2 \omega^2 + \dots,$$

where $a_i \in R$ and all but finitely many a_i are 0. The elements a_i are called **coefficients**, and ω is a formal symbol called an **indeterminate**. The coefficient a_0 is called the **constant term**. If $n \geq 0$ is such that $a_i = 0$ for all $i > n$ and $a_n \neq 0$, then we usually write the above polynomial as

$$a_0 + a_1 \omega + \dots + a_n \omega^n$$

(later we will avoid this lengthy notation and write a polynomial as $f(\omega)$ or simply f). This polynomial is said to have **degree** n, and a_n is called its **leading coefficient**. If $a_i = 0$ for all $i \geq 0$, then the corresponding polynomial is denoted by 0. A polynomial that is either 0 or has degree 0 is called a **constant polynomial**.

The addition and multiplication of two polynomials are defined by

$$\sum_{n=0}^{\infty} a_n \omega^n + \sum_{n=0}^{\infty} b_n \omega^n := \sum_{n=0}^{\infty} (a_n + b_n) \omega^n,$$

$$\left(\sum_{n=0}^{\infty} a_n \omega^n \right) \left(\sum_{n=0}^{\infty} b_n \omega^n \right) := \sum_{n=0}^{\infty} c_n \omega^n, \quad \text{where} \quad c_n = \sum_{i=0}^{n} a_i b_{n-i}.$$

This makes the set of all polynomials over R into a ring, which we denote by $R[\omega]$ and call a **polynomial ring**. If R is unital, then so is $R[\omega]$. Elements in R commute with ω, but they do not necessarily commute with each other. In fact, R is commutative if and only if $R[\omega]$ is commutative. We are primarily interested in the situation where R is a field. In this case we will write R as F and thus denote the polynomial ring as $F[\omega]$.

So far we have considered polynomials in one indeterminate. The notion of a **polynomial in several indeterminates** $\omega_1, \dots, \omega_n$ is not conceptually much more demanding, but notationally heavier. It can be defined as an infinite sum

$$\sum_{i_1=0}^{\infty} \cdots \sum_{i_n=0}^{\infty} a_{i_1,\dots,i_n} \omega_1^{i_1} \dots \omega_n^{i_n},$$

where $a_{i_1,\dots,i_n} \in R$ and all but finitely many $a_{i_1,\dots,i_n}$ are 0. Each term $a_{i_1,\dots,i_n} \omega_1^{i_1} \dots \omega_n^{i_n}$ whose coefficient $a_{i_1,\dots,i_n}$ is not 0 is called a **monomial**. Of course, we may neglect the terms with zero coefficient, and so every polynomial in several indeterminates is a finite sum of monomials. The expressions $\omega_1^{i_1} \dots \omega_n^{i_n}$ basically play the role of components. The addition is defined componentwise. The multiplication is determined by distributivity and the rule

$$(a_{i_1,\dots,i_n} \omega_1^{i_1} \dots \omega_n^{i_n})(b_{j_1,\dots,j_n} \omega_1^{j_1} \dots \omega_n^{j_n}) = (a_{i_1,\dots,i_n} b_{j_1,\dots,j_n}) \omega_1^{i_1+j_1} \dots \omega_n^{i_n+j_n}.$$

These two operations make the set of all polynomials in $\omega_1, \dots, \omega_n$ into a ring which we denote by $R[\omega_1, \dots, \omega_n]$. Incidentally, alternatively one can define $R[\omega_1, \dots, \omega_n]$ recursively as $R[\omega_1, \dots, \omega_{n-1}][\omega_n]$.

One can also start with a possibly infinite set of indeterminates $\Omega = \{\omega_i \mid i \in I\}$, and introduce the polynomial ring $R[\Omega]$. Each polynomial in $R[\Omega]$ involves only finitely many indeterminates, so the above definitions regarding $R[\omega_1, \dots, \omega_n]$ still make sense. Only the notation is even more entangled. Let us point out that each $\omega_i \in \Omega$ is a central element of $R[\Omega]$.

Vector Spaces

A **vector space** over a field F is an additive group V endowed with an additional operation $F \times V \to V$, $(\lambda, v) \mapsto \lambda v$, called the **scalar multiplication**, which satisfies

(a) $(\lambda + \mu)v = \lambda v + \mu v$,
(b) $\lambda(v + w) = \lambda v + \lambda w$,
(c) $\lambda(\mu v) = (\lambda \mu)v$,
(d) $1v = v$

for all $\lambda, \mu \in F$, $v, w \in V$. Note that $\lambda v = 0$ if and only if $\lambda = 0$ or $v = 0$, and $(-\lambda)v = -\lambda v$. Elements in V are called **vectors** and elements in F are called **scalars**. Instead of "vector space" we often say "linear space" or just "space". A vector space over $\mathbb{R}$ is called a **real (vector) space**, and a vector space over $\mathbb{C}$ is called a **complex (vector) space**.

The prototype example of a vector space is F^n, the Cartesian product of n copies of F, endowed with componentwise operations

$$(\lambda_1, \dots, \lambda_n) + (\mu_1, \dots, \mu_n) = (\lambda_1 + \mu_1, \dots, \lambda_n + \mu_n),$$
$$\lambda(\lambda_1, \dots, \lambda_n) = (\lambda\lambda_1, \dots, \lambda\lambda_n).$$

In particular, $F = F^1$ is a vector space. We have introduced $F[\omega]$ and $C[a, b]$ as rings, but they are also vector spaces if we define scalar multiplication in the usual way.

A nonempty subset W of a vector space V over F is a **(linear) subspace** if $\lambda w + \lambda' w' \in W$ for all $\lambda, \lambda' \in F$ and $w, w' \in W$. A subspace is itself a vector space. For example, the set of all polynomials in $F[\omega]$ of degree at most n is a subspace of $F[\omega]$.

A **linear combination** of vectors $v_1, \ldots, v_n$ is every vector of the form $\lambda_1 v_1 + \cdots + \lambda_n v_n$ where $\lambda_i \in F$. If S is a nonempty subset of a space V, then the subspace V_S generated by S is equal to the set of all linear combinations of vectors in S. It is called the **linear span** of S; we also say that V_S is **spanned** by S or that S **spans** V_S. Note that, by definition, a linear combination involves a finite number of vectors. If S is an infinite set, then the linear span of S is the union of the linear spans of all finite subsets of S.

Vector space homomorphisms are **linear maps**, also called **linear operators**. These are maps $\varphi : V \to V'$ that satisfy

$$\varphi(\lambda v + \mu w) = \lambda \varphi(v) + \mu \varphi(w)$$

for all $\lambda, \mu \in F$, $v, w \in V$. The notions such as "kernel", "image", "endomorphism", "isomorphism", etc. should be self-explanatory. Sometimes we will denote linear operators with capital letters (A, B, etc.). A linear map from a vector space over F into the space F is called a **linear functional**. A map $f : V_1 \times V_2 \to V'$, where V_1, V_2, V' are vector spaces over F, is said to be **bilinear** if it is linear in each variable, i.e.,

$$f(\lambda u + \mu v, z) = \lambda f(u, z) + \mu f(v, z) \quad \text{and} \quad f(v, \lambda z + \mu w) = \lambda f(v, z) + \mu f(v, w)$$

for all $\lambda, \mu \in F$, $u, v \in V_1$, $z, w \in V_2$.

A subset S of a vector space V over F is said to be **linearly independent** if for all distinct vectors $s_1, \ldots, s_n \in S$ and all scalars $\lambda_1, \ldots, \lambda_n \in F$, $\lambda_1 s_1 + \cdots + \lambda_n s_n = 0$ implies $\lambda_1 = \cdots = \lambda_n = 0$. If S is not linearly independent, then we say that S is **linearly dependent**. If $S = \{s_i \mid i \in I\}$, then instead of saying that S is linearly (in)dependent we can also say that the vectors s_i, $i \in I$, are linearly (in)dependent.

A linearly independent subset of V that spans V is called a **basis** of V. For example, $\{(1, 0), (1, 1)\}$ is a basis of F^2. The standard basis of F^2 is $\{(1, 0), (0, 1)\}$, and the standard basis of $F[\omega]$ is $\{1, \omega, \omega^2, \ldots\}$. The basis of $\{0\}$ is $\emptyset$.

A vector space V over F is **finite dimensional** if it is finitely generated, i.e., it is spanned by a finite set. Such a space has a finite basis. Moreover, all its bases have the same cardinality, called the **dimension** of V and denoted $\dim_F V$. (The concept of dimension also makes sense for infinite dimensional spaces, but we will not use it for them.) If V is n-dimensional, i.e., $\dim_F V = n$, then every linearly independent set of n vectors in V is a basis of V, every subset of V having more than n vectors is linearly dependent, and the dimension of every proper subspace of V is less than n. An example of such a space is F^n, and in fact every space V with $\dim_F V = n$ is isomorphic to F^n.

We say that a linear operator A from one space to another has **finite rank** if its image is a finite dimensional space. In this case we call the dimension of the image the **rank** of A.

Every vector space, finite or infinite dimensional, has a basis. We will actually prove this in Sect. 3.3, but in a more general framework.

Algebras

A vector space A over a field F is an **algebra** over F, or an F**-algebra**, if it is endowed with associative multiplication $A \times A \to A$, $(x, y) \mapsto xy$, satisfying

$$(\lambda x + \mu y)z = \lambda(xz) + \mu(yz) \quad \text{and} \quad x(\lambda y + \mu z) = \lambda(xy) + \mu(xz)$$

for all $\lambda, \mu \in F$, $x, y, z \in A$. We often say only "algebra" without specifying the associated field. An $\mathbb{R}$-algebra is also called a **real algebra**, and a $\mathbb{C}$-algebra is called a **complex algebra**. There is also the notion of an algebra over a commutative unital ring. The axioms are the same, just the role of a field is replaced by a commutative unital ring. Rings can be then viewed as algebras over $\mathbb{Z}$, which conveniently puts rings and algebras under one roof. However, we will consider only algebras over fields in this book.

Thus, an algebra is a ring which is also a vector space, and multiplication is not only biadditive but bilinear. Many rings mentioned earlier are actually algebras. For example, $F[\omega]$ is an F-algebra and $C[a, b]$ is a real algebra. If A is an algebra, then the ring $M_n(A)$ is an algebra if we define scalar multiplication by

$$\lambda(a_{ij}) := (\lambda a_{ij}).$$

Similarly, the ring $A[\omega]$ becomes an algebra by defining

$$\lambda\left(\sum_{i=0}^{\infty} a_i \omega^i\right) := \sum_{i=0}^{\infty} (\lambda a_i)\omega^i.$$

We say that A is a **finite dimensional algebra** if it is finite dimensional as a vector space. The dimension of A over F will be denoted by

$$[A : F].$$

Thus, $[A : F]$ is nothing but $\dim_F A$, but it will be used only for algebras.

Algebras are, in particular, rings, so all ring-theoretic notions make sense for algebras. However, some need small adjustments. A **subalgebra** is a subring which is also a subspace. Thus, for example, the real algebra $\mathbb{R}$ has no proper nonzero subalgebras, but the ring $\mathbb{R}$ has plenty of subrings. Similarly, **algebra ideals**, one-sided or two-sided, are ideals in the ring-theoretic sense which are also subspaces. If I is an ideal of an F-algebra A, then the factor ring A/I becomes an F-algebra, called the **factor algebra** of A by I, if we define scalar multiplication by

$$\lambda(x + I) := \lambda x + I.$$

The **direct product of algebras** is an algebra with operations defined componentwise. An **algebra homomorphism** is a ring homomorphism which is also a linear map. For example, the aforementioned isomorphism between the ring of matrices of the form $\begin{bmatrix} \alpha & \beta \\ -\beta & \alpha \end{bmatrix}$, $\alpha, \beta \in \mathbb{R}$, and the ring $\mathbb{C}$ is actually an $\mathbb{R}$-algebra isomorphism.

The **center of an algebra** is defined just as the center of a ring. A **unital algebra** is an algebra which is unital as a ring, and a **division algebra** is an algebra which is a division ring. We define the **characteristic of an F-algebra** A, char(A), as the characteristic of A considered as a ring. The axioms readily imply that char(A) coincides with the characteristic of the associated field F as long as A is nonzero; therefore char(A) is either 0 or prime. Hence $\mathbb{Z}_n$ is not an algebra over any field if n is not prime ($\mathbb{Z}_p$, with p prime, is a field and therefore an algebra over itself). It is also easy to see that $\mathbb{Z}$ is not an algebra over any field.

An F-algebra is, in particular, a vector space. Let $\{e_i \mid i \in I\}$ be its basis. For each pair $i, j \in I$ there exist $\alpha_{ijk} \in F$, $k \in I$, such that

$$e_i e_j = \sum_{k \in I} \alpha_{ijk} e_k$$

and all but finitely many α_{ijk}'s are 0. We call these formulas the **multiplication table** (if A has a small dimension, then one can really write this down in the form of a table). Two F-algebras A and B are isomorphic, $A \cong B$, if and only if they have the same multiplication tables with respect to some bases. By this we mean that there exist a basis $\{e_i \mid i \in I\}$ of A, a basis $\{f_i \mid i \in I\}$ of B, and scalars $\alpha_{ijk} \in F$, $i, j, k \in I$, such that $e_i e_j = \sum_{k \in I} \alpha_{ijk} e_k$ and $f_i f_j = \sum_{k \in I} \alpha_{ijk} f_k$ for all $i, j \in I$. Indeed, if this holds then the linear map that sends e_i into f_i is easily seen to be an algebra isomorphism (note that such a map exists and is unique). Conversely, an algebra isomorphism from A onto B clearly maps a basis of A onto a basis of B and preserves the multiplication table. Thus, if $[A : F] = n$, then the multiplication in A is determined by n^3 scalars α_{ijk}. Let us mention that not every choice of the α_{ijk}'s is appropriate since the associativity condition must be fulfilled.

Let A be a nonzero unital algebra. We may consider F as a 1-dimensional subalgebra of A via the canonical embedding $\lambda \mapsto \lambda 1$. Accordingly, the element $\lambda 1 \in A$ will be often written simply as λ. Given $a \in A$ and a polynomial

$$f(\omega) = \lambda_0 + \lambda_1 \omega + \cdots + \lambda_n \omega^n \in F[\omega],$$

we define the **evaluation** of f at a as.

$$f(a) := \lambda_0 + \lambda_1 a + \cdots + \lambda_n a^n \in A.$$

We remark that $f \mapsto f(a)$ is a homomorphism from $F[\omega]$ into A. If f has constant term 0, then we can define $f(a)$ also when A is not unital.

We conclude this section with a particularly important example of an algebra. Let V be a vector space over F. Then $\mathrm{End}_F(V)$, the set of all endomorphisms of V (i.e., linear operators from V into V) becomes an F-algebra if we define addition, scalar multiplication, and multiplication by

$$(A + B)(v) := A(v) + B(v),$$
$$(\lambda A)(v) := \lambda A(v),$$
$$(AB)(v) := A(B(v)).$$

The verification is straightforward. Suppose that $\dim_F V = n < \infty$. Pick a basis $\{e_1, \ldots, e_n\}$ of V. For every $A \in \mathrm{End}_F(V)$ there exist scalars $a_{ij} \in F$ such that

$$A(e_j) = \sum_{i=1}^{n} a_{ij} e_i, \quad j = 1, \ldots, n.$$

A straightforward verification shows that the map $A \mapsto (a_{ij})$ is an algebra isomorphism from $\mathrm{End}_F(V)$ onto the matrix algebra $M_n(F)$ (incidentally, in Sect. 3.4 we will prove a generalization of this fact). Thus,

$$\mathrm{End}_F(V) \cong M_n(F) \text{ if } \dim_F V = n.$$

In particular, $[\mathrm{End}_F(V) : F] = (\dim_F V)^2$.

Fields

We need just a little of field theory in this book. Therefore we will provide the reader only with basic information, but not even all of that will be used.

We start with a field construction which is usually considered as a part of commutative ring theory rather than field theory. A commutative ring without zero-divisors is called a **commutative domain** (perhaps the reader is familiar with a more common term "integral domain", but usually one assumes that integral domains must be unital, which we do not need right now). Typical examples are the ring of integers $\mathbb{Z}$ and the polynomial ring $F[\omega]$ where F is a field. Of course, fields are also commutative domains, but they are too "perfect" to be listed among typical examples.

The field $\mathbb{Q}$ is constructed from the commutative domain $\mathbb{Z}$. Essentially the same construction works for making a field from any nonzero commutative domain Z. Let us sketch it (for more details the reader can look at Sect. 7.1 where this construction is presented in a slightly more general context). Define the relation $\sim$ on the set $Z \times (Z \setminus \{0\})$ by $(r, s) \sim (r', s')$ if $rs' = r's$. It is easy to check that this is an equivalence relation. Denote by rs^{-1} the equivalence class of (r, s). Thus, $rs^{-1} = r's'^{-1}$ if and only if $rs' = r's$. Define addition and multiplication on the set $\widehat{Z}$ of all equivalence classes as follows:

$$rs^{-1} + tu^{-1} := (ru + ts)(su)^{-1},$$
$$rs^{-1} \cdot tu^{-1} := rt(su)^{-1}.$$

These operations are well-defined, and $\widehat{Z}$ is a field. It is called the **field of quotients** of Z. One can embed Z into $\widehat{Z}$ and accordingly consider Z as a subring of $\widehat{Z}$. We remark that there does not exist a subfield E of $\widehat{Z}$ such that $Z \subseteq E \subsetneq \widehat{Z}$. If $Z = \mathbb{Z}$, then $\widehat{Z} = \mathbb{Q}$. If $Z = F[\omega]$, then $\widehat{Z}$ is called the **field of rational functions in** ω and is denoted by $F(\omega)$. More generally, if Ω is an arbitrary set of indeterminates, then the field of quotients of $F[\Omega]$ is called the **field of rational functions in** Ω and denoted by $F(\Omega)$.

We proceed to the genuine field theory. A field K is said to be an **extension field** of a field F if F is a subfield of K. For instance, $\mathbb{R}$ is an extension field of $\mathbb{Q}$, and $\mathbb{C}$ is an extension field of $\mathbb{R}$. Given a field F, one usually wants to construct extension fields satisfying certain requirements. The original field F is often contained in all other fields that appear in the study. In this case we call it the **base field**.

If K is an extension field of F, then we may view K as an algebra over F. Indeed, K is a ring by definition, and the scalar multiplication should be simply understood as the given multiplication in K restricted to the situation where the first factor lies in F—if $\lambda \in F$ and $k \in K$ then $\lambda k \in K$ and all algebra axioms are satisfied. We say that K is a **finite extension** of F if $[K : F] < \infty$. For example, $\mathbb{C}$ is a finite extension of $\mathbb{R}$ with $[\mathbb{C} : \mathbb{R}] = 2$, while $\mathbb{R}$ is not a finite extension of $\mathbb{Q}$. If K is a finite extension of F and L is a finite extension of K, then L is a finite extension of F and $[L : F] = [L : K][K : F]$.

The main point is that K can be viewed as a vector space over F. Let us give a simple illustration of the usefulness of this viewpoint. Let K be a **finite field**. Then $\text{char}(K)$ is a prime number p. Hence K can be considered as a finite dimensional space over the prime subfield $F_0 \cong \mathbb{Z}_p$, and as such is isomorphic to F_0^n for some $n \in \mathbb{N}$. But then $|K| = |F_0^n| = p^n$. We have thereby found out that the cardinality of a finite field can only be p^n with p prime and $n \in \mathbb{N}$. This assertion has a sort of a converse, which, however, is less obvious: For every prime p and $n \in \mathbb{N}$ there exists an (up to isomorphism) unique field having p^n elements.

An element a in an extension field K of F is said to be a **root** of the polynomial $f(\omega) \in F[\omega]$ if $f(a) = 0$. For example, $\sqrt{2} \in \mathbb{R}$ is a root of $\omega^2 - 2 \in \mathbb{Q}[\omega]$. A root of a polynomial of the form $\omega^n - 1$, $n \in \mathbb{N}$, is called a **root of unity**. We say that $a \in K$ is **algebraic** (over F) if a is a root of some nonzero polynomial in $F[\omega]$. More specifically, a is **algebraic of degree** m if it is a root of a polynomial in $F[\omega]$ of degree m, but is not a root of a nonzero polynomial in $F[\omega]$ of degree less than m. In this case the subfield of K generated by a and F is equal to $F[a] := \{\lambda_0 + \lambda_1 a + \cdots + \lambda_{m-1} a^{m-1} \mid \lambda_0, \ldots, \lambda_{m-1} \in F\}$, and $[F[a] : F] = m$. For example, $\mathbb{Q}[\sqrt{2}] = \{\lambda_0 + \lambda_1 \sqrt{2} \mid \lambda_0, \lambda_1 \in \mathbb{Q}\}$.

If $a \in K$ is a root of a polynomial $f(\omega) \in F[\omega]$ of degree n, then there exists a polynomial $q(\omega) \in K[\omega]$ of degree $n - 1$ such that $f(\omega) = (\omega - a)q(\omega)$. From this one easily infers that $f(\omega)$ cannot have more than n roots in K. If $a_1, \ldots, a_n$ are distinct roots of $f(\omega)$ in K, then $f(\omega) = c(\omega - a_1) \ldots (\omega - a_n)$ where $c \in F$ is the leading coefficient of $f(\omega)$.

A field K is said to be **algebraically closed** if every nonconstant polynomial in $K[\omega]$ has a root in K. As is evident from the first assertion in the previous paragraph, this is equivalent to the condition that every polynomial $f(\omega) \in K[\omega]$ of degree $n \geq 1$ splits into linear factors: $f(\omega) = c(\omega - a_1) \ldots (\omega - a_n)$ where $c, a_i \in K$ (some of the a_i's may be equal). The **Fundamental Theorem of Algebra** states that the field $\mathbb{C}$ is algebraically closed. The polynomial $\omega^2 + 1$ does not have a root in $\mathbb{R}$, so $\mathbb{R}$ is not algebraically closed. However, from the fundamental theorem of algebra it can be easily deduced that every polynomial in $\mathbb{R}[\omega]$ splits into linear factors $a + b\omega$ and quadratic factors $c + d\omega + e\omega^2$, where $a, b, c, d, e \in \mathbb{R}$.

Working with algebraically closed fields is often essentially easier than with general fields. Let us only point out one matter that is of importance in noncommutative algebra. Recall that square matrices A and B are **similar** if there exists an invertible matrix T such that $A = TBT^{-1}$. If K is an algebraically closed field, then every matrix $A \in M_n(K)$ is similar to a block diagonal matrix $J \in M_n(K)$ whose blocks are matrices in $M_{k_i}(K)$, $n = \sum_i k_i$, having a fixed scalar λ_i on the main diagonal, the scalar 1 on the superdiagonal, and zeros elsewhere. The matrix J is called the **Jordan normal form** of A. The scalars λ_i appearing in J are the eigenvalues of A.

A field extension $\overline{F}$ of a field F is said to be an **algebraic closure** of F if $\overline{F}$ is algebraically closed and every element in $\overline{F}$ is algebraic over F. For example, $\mathbb{C}$ is an algebraic closure of $\mathbb{R}$. *Every field F has an algebraic closure*, which is, moreover, unique up to isomorphism. This is one of the deeper results stated in this preparatory chapter.

Chapter 1
Finite Dimensional Division Algebras

Finite dimensional division algebras are as simple and "spotless" as a (not necessarily commutative) ring can be. Their definition gathers together all of the most favorable properties. They are also the oldest, the most classical topic of noncommutative algebra. But as it often happens, the simplest objects are not the easiest to study, and classical themes are not less profound than modern ones. In many algebra textbooks finite dimensional division algebras are treated in the last chapters by using advanced tools. Our goals in this first chapter, however, are relatively modest. We will give an elementary, self-contained introduction to finite dimensional algebras, which in particular includes two results of extreme importance and beauty: Frobenius' theorem on real division algebras and Wedderburn's theorem on finite division rings. We shall not strictly confine ourselves to division algebras; presenting some of the results in the context of (central) simple algebras, even not necessarily finite dimensional ones, will not change the level of complexity of arguments and shall prove useful later.

1.1 After the Complex Numbers: What Comes Next?

A 2-dimensional real vector space can be transformed into the complex numbers by appropriately defining multiplication. One just takes two linearly independent vectors, denotes one of them by 1 to indicate that it plays the role of a unity, requires that the square of the other one is -1, and then extends this multiplication to the whole space by bilinearity. If one is interested in the geometric interpretation of this multiplication, then one will choose these two vectors more carefully. However, from the algebraic point of view it does not matter which two vectors we take, as long as they are linearly independent and therefore form a basis of our space. In any case we get a 2-dimensional $\mathbb{R}$-algebra isomorphic to the complex number field $\mathbb{C}$.

Why confine ourselves to 2-dimensional spaces? Are there other ways to create some kind of "numbers" from finite dimensional real vector spaces? In any set of "numbers" one should be able to add, subtract, and multiply elements, and multiplication should be associative and bilinear. Thus, our space should in particular

© Springer International Publishing Switzerland 2014
M. Brešar, *Introduction to Noncommutative Algebra*, Universitext,
DOI 10.1007/978-3-319-08693-4_1

be an $\mathbb{R}$-algebra (in the early literature elements from algebras were actually called *hypercomplex numbers*). In a decent set of "numbers" one should also be able to divide elements, with the obvious exclusion of dividing by 0. We can now rephrase our question in a rigorous manner as follows: Is it possible to define multiplication on an n-dimensional real space so that it becomes a real division algebra?

For $n = 1$ the question is trivial; every element is a scalar multiple of unity and therefore up to isomorphism $\mathbb{R}$ itself is the only such algebra. For $n = 2$ we know one example, $\mathbb{C}$, but are there any other? This question is quite easy and the reader may try to solve it immediately. What about $n = 3$? This is already a nontrivial question, but certainly a natural and challenging one. After finding a multiplication in 2-dimensional spaces of such enormous importance in mathematics, should not the next step be finding something similar in 3-dimensional spaces? Furthermore, what about $n = 4, 5, 6, \ldots$?

We will give complete answers to the above questions. Throughout the section, we *assume that D is an n-dimensional division $\mathbb{R}$-algebra*. We will find all possible n for which such an algebra exists, and moreover, all possible multiplications making a finite dimensional real space into a division algebra.

First we recall the notational convention. For $\lambda \in \mathbb{R}$, we write $\lambda 1 \in D$ simply as λ. In fact we identify $\mathbb{R}$ with $\mathbb{R}1$, and in this way consider $\mathbb{R}$ as a subalgebra of D.

Lemma 1.1 *For every $x \in D$ there exists $\lambda \in \mathbb{R}$ such that $x^2 + \lambda x \in \mathbb{R}$.*

Proof Since the dimension of D is n, the elements $1, x, \ldots, x^n$ are linearly dependent. This means that there exists a nonzero polynomial $f(\omega) \in \mathbb{R}[\omega]$ of degree at most n such that $f(x) = 0$. We may assume that the leading coefficient of $f(\omega)$ is equal to 1. As we know, $f(\omega)$ splits into linear and quadratic factors in $\mathbb{R}[\omega]$:

$$f(\omega) = (\omega - \alpha_1) \ldots (\omega - \alpha_r)(\omega^2 + \lambda_1 \omega + \mu_1) \ldots (\omega^2 + \lambda_s \omega + \mu_s)$$

where $\alpha_i, \lambda_i, \mu_i \in \mathbb{R}$. Since $f(x) = 0$, we have

$$(x - \alpha_1) \ldots (x - \alpha_r)(x^2 + \lambda_1 x + \mu_1) \ldots (x^2 + \lambda_s x + \mu_s) = 0.$$

As D is a division algebra, one of the factors must be 0. That is, x is a root of a linear or quadratic polynomial in $\mathbb{R}[\omega]$. In any case the desired conclusion holds.   □

The 1-dimensional subspace $\mathbb{R}i$ of $\mathbb{C}$ can be described as the set of all $z \in \mathbb{C}$ such that z^2 is a nonpositive real number. The analysis of the set of all such elements in our abstract setting will lead us to the answers to our questions.

Lemma 1.2 *The set $V = \{v \in D \mid v^2 \in \mathbb{R}, v^2 \le 0\}$ is a linear subspace of D. Moreover, $D = \mathbb{R} \oplus V$.*

Proof We start with a useful observation. If $x \in D \backslash V$ is such that $x^2 \in \mathbb{R}$, then $x^2 > 0$ and so $x^2 = \alpha^2$ for some $\alpha \in \mathbb{R}$. Thus $(x - \alpha)(x + \alpha) = 0$, giving $x = \pm \alpha \in \mathbb{R}$.

It is clear that $\mathbb{R} \cap V = 0$ and that V is closed under scalar multiplication. Let us show that $u + v \in V$ if $u, v \in V$. We may assume that u and v are linearly independent. We claim that $u, v, 1$ are also independent. Indeed, taking $\alpha, \beta, \gamma \in \mathbb{R}$ such that $\alpha u = \beta v + \gamma$ it follows by squaring both sides that $\beta \gamma v \in \mathbb{R}$, hence $\beta = 0$ or $\gamma = 0$, and $\alpha = \beta = \gamma = 0$ now readily follows. By Lemma 1.1 there exist $\lambda, \mu \in \mathbb{R}$ such that

$$(u + v)^2 + \lambda(u + v) \in \mathbb{R} \text{ and } (u - v)^2 + \mu(u - v) \in \mathbb{R}.$$

On the other hand,

$$(u + v)^2 + (u - v)^2 = 2u^2 + 2v^2 \in \mathbb{R}.$$

Comparing all the relations we obtain $\lambda(u + v) + \mu(u - v) \in \mathbb{R}$. Since $u, v, 1$ are linearly independent, this yields $\lambda + \mu = \lambda - \mu = 0$, hence $\lambda = \mu = 0$, and $u + v \in V$ follows from the first paragraph. Thus, V is a subspace of D. If $x \in D \backslash \mathbb{R}$, then, by Lemma 1.1, $x^2 + vx \in \mathbb{R}$ for some $v \in \mathbb{R}$. Again using the observation from the first paragraph we see that $x + \frac{v}{2} \in V$. Accordingly, $x = -\frac{v}{2} + (x + \frac{v}{2}) \in \mathbb{R} \oplus V$. $\qquad\square$

We set $u \circ v := uv + vu$ for $u, v \in V$. Since $u \circ v = (u + v)^2 - u^2 - v^2$ it follows from Lemma 1.2 that $u \circ v \in \mathbb{R}$. If $v \neq 0$, then $v \circ v = 2v^2 \neq 0$.

Lemma 1.3 *If $n > 2$, then there exist $i, j, k \in D$ such that*

$$i^2 = j^2 = k^2 = -1,$$
$$ij = -ji = k, \quad ki = -ik = j, \quad jk = -kj = i, \tag{1.1}$$

and $1, i, j, k$ are linearly independent.

Proof Lemma 1.2 shows that V has dimension $n - 1 > 1$. Therefore we may choose linearly independent $v, w \in V$. Let $u := w - \frac{w \circ v}{v \circ v} v$. Observe that $u \neq 0$ and $u \circ v = 0$. Set $i := \frac{1}{\sqrt{-u^2}} u, j := \frac{1}{\sqrt{-v^2}} v$, and $k := ij$. It is straightforward to check that (1.1) holds, from which we obtain that

$$(\alpha_0 + \alpha_1 i + \alpha_2 j + \alpha_3 k)(\alpha_0 - \alpha_1 i - \alpha_2 j - \alpha_3 k) = \alpha_0^2 + \alpha_1^2 + \alpha_2^2 + \alpha_3^2 \tag{1.2}$$

is a positive real number for any $\alpha_0, \alpha_1, \alpha_2, \alpha_3 \in \mathbb{R}$ that are not all 0. This proves, in particular, that $1, i, j, k$ are linearly independent. $\qquad\square$

The second formula in (1.1) can be easily remembered by the following picture:

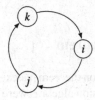

Going clockwise, the product of two consecutive elements is the third one; going counterclockwise, we obtain the negative of the third one.

Lemma 1.3 rules out the case where $n = 3$. If $n = 4$ then D has a basis $1, i, j, k$ with multiplication table (1.1). This is a real finding. Let us change the point of view. Suppose we are given a 4-dimensional real vector space D with basis $\{1, i, j, k\}$ in which we define multiplication by (1.1) and the requirement that 1 is a unity. A direct verification shows that D is indeed an $\mathbb{R}$-algebra (note that it is enough to check associativity for elements from a basis!). We define the **conjugate** of an arbitrary element $h = \alpha_0 + \alpha_1 i + \alpha_2 j + \alpha_3 k$ in D by

$$\overline{h} := \alpha_0 - \alpha_1 i - \alpha_2 j - \alpha_3 k.$$

If $h \neq 0$, then $h\overline{h} (= \overline{h}h)$ is a nonzero scalar by (1.2). Therefore h is invertible with inverse $\frac{1}{h\overline{h}}\overline{h}$. Accordingly, D is a division algebra.

The standard notation for this 4-dimensional noncommutative real division algebra is $\mathbb{H}$, in honor of its discoverer, the Irish mathematician W. R. Hamilton. The elements of $\mathbb{H}$ are called (real) **quaternions**. As the first noncommutative algebraic system ever introduced, $\mathbb{H}$ is of great historic value. Before this discovery Hamilton had been struggling for a long time to extend the multiplication of complex numbers to the 3-dimensional space, that is, he had wanted to solve our problem for $n = 3$. We have just seen that this had been hopeless. On October 16, 1843, when walking with his wife along the Royal Canal in Dublin, the idea of 4-dimensional quaternions suddenly came into his mind, and he carved the formula $i^2 = j^2 = k^2 = ijk = -1$ in a stone of Broom Bridge. This story may have a flavor of a romantic legend, but is documented in Hamilton's letters.

The following theorem was proved in 1878 by another great mathematician, F. G. Frobenius.

Theorem 1.4 (Frobenius) *A finite dimensional real division algebra D is isomorphic to $\mathbb{R}$, $\mathbb{C}$, or $\mathbb{H}$.*

Proof If $n = 1$, then $D \cong \mathbb{R}$. If $n = 2$, then $V \neq 0$ by Lemma 1.2, so D contains an element i such that $i^2 = -1$; but then $D \cong \mathbb{C}$. By Lemma 1.3 we know that $n \neq 3$ and that $D \cong \mathbb{H}$ if $n = 4$.

Suppose $n > 4$. Let i, j, k be the elements from Lemma 1.3. Since the dimension of V is $n - 1$, there exists $v \in V$ not lying in the linear span of i, j, k. Therefore $e := v + \frac{iov}{2}i + \frac{jov}{2}j + \frac{kov}{2}k$ is a nonzero element in V and it satisfies $i \circ e = j \circ e = k \circ e = 0$. However, from the first two identities we conclude $eij = -iej = ije$, which contradicts the third identity since $ij = k$. Thus, n can only be 1, 2, or 4. $\square$

1.2 Beyond Frobenius' Theorem

Frobenius' theorem is a classification theorem par excellence. Three important examples of finite dimensional real division algebras were discovered by mathematicians, each of them has turned out to be immensely important, and, up to isomorphism, these three are in fact the only ones that exist. Yet this definitive result has various extensions. However, they are mostly concerned with **nonassociative** algebras, i.e., algebras in which multiplication is not necessarily associative. This goes beyond the scope of this book. Still, at this point it seems appropriate to at least mention the algebra of **octonions** $\mathbb{O}$, the natural successor in the sequence $\mathbb{R}, \mathbb{C}, \mathbb{H}$. This is an 8-dimensional nonassociative real algebra which contains a copy of $\mathbb{H}$ as its subalgebra. Its standard basis contains a unity 1 and seven elements whose squares equal -1. Although not associative, $\mathbb{O}$ satisfies weak associativity conditions $x^2 y = x(xy)$ and $yx^2 = (yx)x$ for all $x, y \in \mathbb{O}$; nonassociative algebras satisfying these two conditions are called **alternative**. Further, $\mathbb{O}$ is a **nonassociative division algebra**, which, by definition, means that for all $a, b \in \mathbb{O}$ with $a \neq 0$, the equations $ax = b$ and $ya = b$ have unique solutions in $\mathbb{O}$. Let this be enough in our attempt to make the reader curious and challenged for seeking further information about octonions elsewhere. Two books that include an introduction to nonassociative algebras and can be recommended are [McC04] and [ZSSS82].

Let us return to our theme. A close inspection of the arguments from the previous section gives further insight. First, does the last paragraph of the proof of Theorem 1.4 resemble something familiar? Observe that $(u, v) \mapsto -\frac{1}{2} u \circ v$ defines an inner product on V such that i, j, k are orthonormal vectors, so we were performing nothing but the Gram-Schmidt process when introducing e. Moreover, this is not just any inner product. Identifying each $u = \alpha_1 i + \alpha_2 j + \alpha_3 k \in V$ with $\mathbf{u} = (\alpha_1, \alpha_2, \alpha_3) \in \mathbb{R}^3$, we have $-\frac{1}{2} u \circ v = \mathbf{u} \cdot \mathbf{v}$ where $\cdot$ is the dot product. Another interesting observation is that $\frac{1}{2}(uv - vu) = \mathbf{u} \times \mathbf{v}$ where $\times$ is the cross product. Using $uv = \frac{1}{2} u \circ v + \frac{1}{2}(uv - vu)$ one now easily sees that $\mathbb{H}$ can be equivalently introduced as the vector space $\mathbb{R} \times \mathbb{R}^3$ (with pointwise addition and scalar multiplication) endowed with multiplication

$$(\alpha_0, \mathbf{u})(\beta_0, \mathbf{v}) = (\alpha_0 \beta_0 - \mathbf{u} \cdot \mathbf{v}, \alpha_0 \mathbf{v} + \beta_0 \mathbf{u} + \mathbf{u} \times \mathbf{v}).$$

Let us mention in passing that another standard way is to introduce quaternions as certain 2×2 complex matrices. We will arrive at this in Example 2.72.

Our second remark concerning the previous section is that the finite dimensionality of D was used only at one point, namely at the beginning of the proof of Lemma 1.1 when concluding that every $x \in D$ is a root of a nonzero polynomial in $\mathbb{R}[\omega]$. This observation yields a slightly different interpretation of the obtained result. Let us first introduce some terminology. We say that an element x from an F-algebra A is **algebraic** if there exists a nonzero polynomial $f(\omega) \in F[\omega]$ such that $f(x) = 0$. This is, of course, just the repetition of the definition given in Prerequisites for field extensions, and just as before we define the **degree of algebraicity**. If every element in A is algebraic, then A is said to be an **algebraic algebra**. The following simple, fundamental fact was indirectly observed in the proof of Lemma 1.1.

Lemma 1.5 *Every finite dimensional algebra is an algebraic algebra.*

Proof The set of all powers of each element is linearly dependent. $\square$

Our proof of Frobenius' theorem thus shows that an algebraic real division algebra is isomorphic to $\mathbb{R}$, $\mathbb{C}$, or $\mathbb{H}$, and hence it is finite dimensional. Let us mention that infinite dimensional algebraic algebras (over any field) do exist, and will also be met in this book (for the first time in Example 2.7).

After describing finite dimensional real division algebras, one may wonder what about the complex ones? The answer is simple and comes easily. Instead of $\mathbb{C}$ we can actually take any algebraically closed field.

Proposition 1.6 *If D is a finite dimensional division algebra over an algebraically closed field F, then $D = F$.*

Proof Let $x \in D$. By Lemma 1.5 there exists a nonzero polynomial $f(\omega) \in F[\omega]$ such that $f(x) = 0$. We may assume that its leading coefficient equals 1. Since F is algebraically closed, we have $f(\omega) = (\omega - \alpha_1)\dots(\omega - \alpha_r)$ for some $\alpha_i \in F$. Consequently, $(x - \alpha_1)\dots(x - \alpha_r) = 0$. As D is a division algebra, this gives $x - \alpha_i = 0$ for some i, and hence $x = \alpha_i \in F$. $\square$

Of course, again we may substitute "algebraic" for "finite dimensional".

This proposition may seem disappointing after encountering the quaternions. But later we will see that it can be actually considered as "good news".

Remark 1.7 The 4-dimensional algebra with basis $\{1, i, j, k\}$ and multiplication determined by (1.1) can be formed over any field F, not only over $\mathbb{R}$. However, if $F = \mathbb{C}$, then this is not a division algebra. In fact, from (1.2) we see that it is a division algebra if and only if for all $\alpha_i \in F$, $\alpha_0^2 + \alpha_1^2 + \alpha_2^2 + \alpha_3^2 = 0$ implies $\alpha_0 = \alpha_1 = \alpha_2 = \alpha_3 = 0$. This condition is of course fulfilled in $\mathbb{R}$ and some other fields (say, in $\mathbb{Q}$), but not in algebraically closed fields. Still, one might wonder what algebra we get if $F = \mathbb{C}$. This is not a hard question, but we will answer it later, in Example 2.72.

1.3 Simple Rings

Our next main goal is a theorem on finite division rings. In the next sections we shall progress slowly towards its proof, making several digressions that will turn out to be important later when dealing with more general rings and algebras. For a while division algebras will mostly lie dormant in our exposition. This section is devoted to the following notion.

Definition 1.8 A ring R is said to be **simple** if $R^2 \neq 0$ and 0 and R are the only ideals of R.

The condition that $R^2 \neq 0$, i.e., $xy \neq 0$ for some $x, y \in R$, is needed to exclude pathological cases. In nonzero unital rings it is automatically fulfilled. The condition about ideals is, of course, more restrictive. Most of rings are not simple, yet some of the most significant are. At any rate, simple rings form one of the most important classes of rings.

Example 1.9 Every division ring D is simple. Moreover, D does not even have one-sided ideals different from 0 and D. This follows immediately from the elementary observation that proper one-sided ideals cannot contain invertible elements.

There are many important simple rings that are not division rings. We will now present the most fundamental example. First we have to introduce some notation and terminology.

Consider the matrix ring $M_n(S)$ where S is an arbitrary unital ring. Let E_{ij} denote the matrix whose (i,j) entry is 1 and all other entries are 0. We call E_{ij}, $1 \le i,j \le n$, the **standard matrix units**. By aE_{ij} we denote the matrix whose (i,j) entry is $a \in S$ and other entries are 0. Every matrix in $M_n(S)$ is a sum of such matrices. Note that for every $A = (a_{ij}) \in M_n(S)$ we have

$$E_{ij}AE_{kl} = a_{jk}E_{il} \quad \text{for all } 1 \le i,j,k,l \le n. \tag{1.3}$$

This formula considerably simplifies computations in $M_n(S)$. Yet the reader might be more persuaded in the correctness of the arguments based on (1.3) by occasionally writing down matrices and then making the usual matrix calculations. Just do this with 2×2 matrices, it is enough to feel certain.

Example 1.10 If D is a division ring, then $M_n(D)$ is a simple ring for every $n \in \mathbb{N}$. Indeed, let I be a nonzero ideal of $M_n(D)$. Pick a nonzero $(a_{ij}) \in I$. Choose j and k so that $a_{jk} \ne 0$. From (1.3) we see that $a_{jk}E_{il} \in I$ for all i and l, and hence also $(da_{jk}^{-1})E_{ii} \cdot a_{jk}E_{il} = dE_{il} \in I$ for every $d \in D$. Consequently, $I = M_n(D)$.

Remark 1.11 Rather than in simple rings, we are currently interested in **simple algebras**, i.e., algebras satisfying the conditions of Definition 1.8. This may sound a bit ambiguous. Namely, an ideal of an algebra A must be a vector subspace, while Definition 1.8 speaks about ideals of a ring which are merely additive subgroups. Fortunately, both possible interpretations lead to the same conclusion. Indeed, assume that A is an algebra such that 0 and A are the only algebra ideals of A, and I is a ring ideal of A such that $I \ne 0$ and $I \ne A$. Then AI is an algebra ideal of A, so it is either 0 or A. Since $AI \subseteq I \ne A$, the only possibility is that $AI = 0$. This implies that the algebra ideal $J = \{x \in A \mid Ax = 0\}$ is nonzero, so that $J = A$. That is, $A^2 = 0$. This possibility is excluded in Definition 1.8. Thus, an algebra A is simple as an algebra if and only if it is simple as a ring.

Without assuming that $A^2 \ne 0$ it is possible that A has many ring ideals, but no algebra ideals different from 0 and A. For instance, take the 1-dimensional algebra $A = \mathbb{R}a$ with trivial multiplication, i.e., $a^2 = 0$. Then clearly A has no algebra ideals different from 0 and A, but has plenty of ring ideals (say, $\mathbb{Z}a$).

Our next goal is to give an example of an infinite dimensional simple algebra which will accompany us throughout the book. First we introduce a useful notion that naturally arises in noncommutative algebra.

Definition 1.12 The **commutator** of elements a and b in a ring R is the element

$$[a, b] := ab - ba.$$

Obviously, a and b commute if and only if $[a, b] = 0$. We also remark that

$$[a, b] = -[b, a] \quad \text{and} \quad [a, bc] = [a, b]c + b[a, c].$$

Example 1.13 Let F be a field with $\text{char}(F) = 0$. Recall that $\text{End}_F(F[\omega])$ denotes the algebra of all linear operators of the vector space $F[\omega]$. Define $D, L \in \text{End}_F(F[\omega])$ by

$$D\big(f(\omega)\big) = f'(\omega), \quad L\big(f(\omega)\big) = \omega f(\omega).$$

Here, $f'(\omega)$ is the derivative of $f(\omega)$. The subalgebra of $\text{End}_F(F[\omega])$ generated by D and L is called the **Weyl algebra**. We denote it by $\mathscr{A}_1$. A more accurate name is **the first Weyl algebra**. We will introduce the higher Weyl algebras $\mathscr{A}_n$ later, in Example 6.4. The origin of these algebras is in quantum mechanics, but we will see that they are also very interesting algebraic objects. Let us examine some properties of $\mathscr{A}_1$.

It is straightforward to check that

$$[D, L] = I, \tag{1.4}$$

the identity operator. This is the basic relation in $\mathscr{A}_1$. Everything what follows will be deduced from it. In particular we now see that $\mathscr{A}_1$ is a unital algebra. Next, by induction on n one easily shows that

$$[D, L^n] = [D, L^{n-1}]L + L^{n-1}[D, L] = nL^{n-1} \tag{1.5}$$

for every $n \in \mathbb{N}$ (here we follow the convention that $T^0 = I$ for every operator T). Similarly we get

$$[D^n, L] = nD^{n-1}. \tag{1.6}$$

Let $\mathscr{L}$ be the subalgebra of $\mathscr{A}_1$ generated by I and L. From (1.5) we see that $[D, \mathscr{L}] \subseteq \mathscr{L}$; in particular,

$$D\mathscr{L} \subseteq \mathscr{L} + \mathscr{L}D. \tag{1.7}$$

Using $[D^m, T] = D[D^{m-1}, T] + [D, T]D^{m-1}$ we obtain inductively that

$$[D^m, \mathscr{L}] \subseteq \mathscr{L} + \mathscr{L}D + \cdots + \mathscr{L}D^{m-1} \quad \text{for every } m \in \mathbb{N}. \tag{1.8}$$

This readily implies that the linear span of $\{TD^m \mid T \in \mathscr{L}, m \geq 0\}$ is a subalgebra of $\mathscr{A}_1$, and hence, since it contains L and D, it is equal to $\mathscr{A}_1$. That is, we have shown that

(a) *Every element in $\mathscr{A}_1$ can be written as $T_0 + T_1D + \cdots + T_nD^n$, where $n \geq 0$ and $T_0, T_1, \ldots, T_n \in \mathscr{L}$.*

We claim that the elements T_i are uniquely determined. Thus, we have to show that $T_0 + T_1D + \cdots + T_nD^n = 0$ implies $T_i = 0$ for each i. And this is easy. Considering the action of the left-hand side operator on 1 we get $T_0(1) = 0$, and hence $T_0 = 0$. Similarly, by considering the action on ω we get $T_1 = 0$. We continue with ω^2, ω^3, etc. Note that the assumption that $\mathrm{char}(F) = 0$ is used when deriving $T_2 = 0$, $T_3 = 0$, etc. Since $\{L^m \mid m \geq 0\}$ is obviously a basis of $\mathscr{L}$, we have thereby proved that

(b) *The set $\{L^mD^n \mid m, n \geq 0\}$ is a basis of $\mathscr{A}_1$.*

(Analogously one shows that so is the set $\{D^mL^n \mid m, n \geq 0\}$.)
    Let us finally prove that

(c) *$\mathscr{A}_1$ is a simple algebra.*

Let $\mathscr{I}$ be a nonzero ideal of $\mathscr{A}_1$. Take $S = \sum_{k=0}^{n} T_kD^k \in \mathscr{I}$, $T_k \in \mathscr{L}$, such that $T_n \neq 0$ and n is minimal. Suppose $n > 0$. Since L commutes with each T_i, it follows, by making use of (1.6), that

$$[S, L] = \sum_{k=0}^{n'} T_k[D^k, L] = \sum_{k=1}^{n} kT_kD^{k-1}. \tag{1.9}$$

However, since $nT_n \neq 0$ and $[S, L] \in \mathscr{I}$, this contradicts the minimality of n. Thus $n = 0$, i.e., $\mathscr{I} \cap \mathscr{L} \neq 0$. Now take $T = \sum_{j=0}^{m} \alpha_jL^j \in \mathscr{I} \cap \mathscr{L}$, $\alpha_j \in F$, such that $\alpha_m \neq 0$ and m is minimal. Suppose $m > 0$. Using (1.5) we obtain

$$[D, T] = \sum_{j=1}^{m} j\alpha_jL^{j-1}, \tag{1.10}$$

contradicting the minimality of m. Therefore $m = 0$, i.e., $I \in \mathscr{I}$, and so $\mathscr{I} = \mathscr{A}_1$. This proves (c).

1.4 Central Algebras

The center of a unital algebra obviously contains scalar multiples of unity. In many important examples of algebras these are also the only central elements.

Definition 1.14 A nonzero unital algebra is said to be **central** if scalar multiples of unity are the only elements in its center.

    Unless specified otherwise, by an algebra we shall mean an algebra over the field F. If A is a nonzero unital algebra, then we identify F with $F \cdot 1$, and write λ instead of $\lambda 1$ for every scalar $\lambda \in F$. Thus, A is central if $Z(A) = F$. In other words, A is central if it is an algebra over its center.

Lemma 1.15 *Let A be a unital algebra and let $n \in \mathbb{N}$. Then A is central if and only if $M_n(A)$ is central.*

Proof If $c \in Z(A)$, then the diagonal matrix with all diagonal entries equal to c lies in $Z(M_n(A))$. Thus, A is central if $M_n(A)$ is. Now let A be central. Pick $C \in Z(M_n(A))$. Since C commutes with every E_{kl}, $k \neq l$, one easily shows that C is a diagonal matrix with all diagonal entries equal. As C also commutes with every $aE_{11}, a \in A$, we conclude that this diagonal entry lies in the center of A, and hence it is a scalar. Thus $M_n(A)$ is central. $\square$

Example 1.16 As the simplest application of Lemma 1.15 we have that $M_n(F)$ is a central algebra.

Example 1.17 It is easy to verify that $\mathbb{H}$ is a central $\mathbb{R}$-algebra.

Example 1.18 Obviously, $\mathbb{C}$ is central as a $\mathbb{C}$-algebra, but not as an $\mathbb{R}$-algebra.

Theorem 1.4 therefore implies

Corollary 1.19 *A finite dimensional central division $\mathbb{R}$-algebra is isomorphic to $\mathbb{R}$ or $\mathbb{H}$.*

Example 1.20 The Weyl algebra $\mathscr{A}_1$ is central. Indeed, take $S = \sum_{k=0}^{n} T_k D^k \in Z(\mathscr{A}_1)$. As $[S, L] = 0$, it follows from (1.9) that $T_i = 0$ for $i \geq 1$. Therefore $S = \sum_{j=0}^{m} \alpha_j L^j$. Using $[D, S] = 0$ we see from (1.10) that $\alpha_i = 0$ for $i \geq 1$. Thus $S = \alpha_0 \in F$.

Example 1.21 If A is a simple unital ring, then its center is a field. Indeed, if c is a nonzero central element, then cA must be, as a nonzero ideal of A, equal to A. This implies that c is invertible. If we multiply $cx = xc$ from the left and right by c^{-1} we see that c^{-1} is also a central element. Consequently, *every simple unital ring can be viewed as a central algebra over its center.* Namely, we can define scalar multiplication simply as the ordinary multiplication of a central element with an arbitrary element from A, and all algebra axioms are readily fulfilled.

The main attention in the following sections will be devoted to *finite dimensional central simple algebras,* which form an extremely important class of algebras. For instance, the algebras $M_n(F)$ and $\mathbb{H}$ belong to this class, as it is evident from the above examples. Later we will see that every finite dimensional central simple algebra is built from matrices and division algebras (Corollary 2.62). However, our immediate goal is to derive some results on these algebras directly from the definition, without relying on the description of their structure. For this purpose we will now introduce a certain algebra of linear operators.

1.5 Multiplication Algebra

Let A be an algebra. For $a, b \in A$ we define maps $L_a, R_b : A \to A$, called the **left multiplication map** and **right multiplication map**, by

$$L_a(x) := ax, \quad R_b(x) := xb.$$

These maps will frequently play important roles in this book. Of course, they can be defined if A is merely a ring, but we will usually treat them on algebras. In this case they can be considered as elements of $\mathrm{End}_F(A)$, the algebra of all linear operators from A into itself.

Note that for all $a, b \in A$, $\lambda, \mu \in F$ we have

$$L_{ab} = L_a L_b, \quad R_{ab} = R_b R_a,$$
$$L_a R_b = R_b L_a,$$
$$L_{\lambda a + \mu b} = \lambda L_a + \mu L_b, \quad R_{\lambda a + \mu b} = \lambda R_a + \mu R_b.$$

Hence we conclude that

$$M(A) := \{L_{a_1} R_{b_1} + \cdots + L_{a_n} R_{b_n} \mid a_i, b_i \in A, n \in \mathbb{N}\}$$

is a subalgebra of $\mathrm{End}_F(A)$.

Definition 1.22 The algebra $M(A)$ is called the **multiplication algebra of A**.

If A is unital, then $L_a = L_a R_1$ and $R_b = L_1 R_b$ lie in $M(A)$, and so in this case we can describe $M(A)$ as the subalgebra of $\mathrm{End}_F(A)$ generated by all left and right multiplication maps. We also remark that $a = L_a(1) = R_a(1)$, and so the conditions $a = 0$, $L_a = 0$, and $R_a = 0$ are equivalent if A is unital.

Remark 1.23 Take $f \in M(A)$. Then there exist $a_i, b_i \in A$ such that $f = \sum_{i=1}^{n} L_{a_i} R_{b_i}$, i.e.,

$$f(x) = \sum_{i=1}^{n} a_i x b_i, \quad x \in A.$$

These elements are not unique, f can be presented as a sum of operators of the form $L_a R_b$ in many ways. If $f \neq 0$, then a_i, b_i can be chosen so that both the set of all a_i's and the set of all b_i's are linearly independent. For example, this is fulfilled if we require that n is minimal. Indeed, if, under this assumption, one of the elements, say b_n, was a linear combination of the others, $b_n = \sum_{i=1}^{n-1} \lambda_i b_i$, then f would be equal to

$$\sum_{i=1}^{n-1} L_{a_i + \lambda_i a_n} R_{b_i},$$

contradicting the minimality of n. Furthermore, if $\{u_i \mid i \in I\}$ is a basis of A, then we can express each R_b as a linear combination of R_{u_i}, from which we see that every $f \in M(A)$ is a (finite) sum of operators of the form $L_{a_i}R_{u_i}$. Similarly, f can be written as a sum of operators of the form $L_{u_i}R_{b_i}$, and, moreover, as a linear combination of operators $L_{u_i}R_{u_j}$.

The fact that the elements from $M(A)$ can be expressed in different ways through left and right multiplication maps may create some annoyance. The following lemma shows that in central simple algebras this problem is controllable.

Lemma 1.24 *Let A be a central simple algebra, and let $a_i, b_i \in A$ be such that $\sum_{i=1}^{n} L_{a_i}R_{b_i} = 0$. If the a_i's are linearly independent, then each $b_i = 0$. Similarly, if the b_i's are linearly independent, then each $a_i = 0$.*

Proof The two assertions of the lemma are analogous, so we consider only the case where the a_i's are independent. Suppose $b_n \neq 0$. Since A is simple, the ideal generated by b_n is equal to A. That is, $\sum_{j=1}^{m} w_j b_n z_j = 1$ for some $w_j, z_j \in A$. Hence

$$0 = \sum_{j=1}^{m} R_{z_j}\left(\sum_{i=1}^{n} L_{a_i}R_{b_i}\right)R_{w_j} = \sum_{i=1}^{n} L_{a_i}\left(\sum_{j=1}^{m} R_{w_j b_i z_j}\right) = \sum_{i=1}^{n} L_{a_i}R_{c_i}$$

where $c_i = \sum_{j=1}^{m} w_j b_i z_j$; thus, $c_n = 1$. This clearly implies that $n > 1$. We may assume that n is the smallest natural number for which the lemma does not hold. Since

$$0 = \left(\sum_{i=1}^{n} L_{a_i}R_{c_i}\right)R_x - R_x\left(\sum_{i=1}^{n} L_{a_i}R_{c_i}\right) = \sum_{i=1}^{n-1} L_{a_i}R_{x c_i - c_i x}$$

for every $x \in A$, it thus follows that $x c_i - c_i x = 0$. Consequently, $c_i \in F$. But then

$$0 = \sum_{i=1}^{n} L_{a_i}R_{c_i} = L_{c_1 a_1 + \cdots + c_n a_n},$$

which contradicts the linear independence of the a_i's.   □

In the next lemma we consider the finite dimensional situation. Let us recall that if $[A : F] = d$, then $[\mathrm{End}_F(A) : F] = d^2$.

Lemma 1.25 *If A is a finite dimensional central simple algebra, then $M(A) = \mathrm{End}_F(A)$.*

Proof Let $\{u_1, \ldots, u_d\}$ be a basis of A. Lemma 1.24 implies that the operators $L_{u_i}R_{u_j}$, $1 \leq i, j \leq d$, are linearly independent. This becomes clear if we rewrite $\sum_{i,j=1}^{d} \lambda_{ij} L_{u_i}R_{u_j}$ as $\sum_{i=1}^{d} L_{u_i}R_{b_i}$ where $b_i = \sum_{j=1}^{d} \lambda_{ij} u_j$. Therefore

$$[M(A) : F] \geq d^2 = [\text{End}_F(A) : F],$$

and so $M(A) = \text{End}_F(A)$.   □

It may be instructive for the reader to find an independent, direct proof of this lemma for the special case where $A = M_n(F)$ (what is a natural basis of $\text{End}_F(A)$?).

Lemmas 1.24 and 1.25 will be used at different places throughout the book. In the next sections we will see that three classical results can be deduced from them.

1.6 Automorphisms of Central Simple Algebras

The first result that we will derive from Lemmas 1.24 and 1.25 is an important special case of a celebrated theorem established by T. Skolem in 1927 and somewhat later rediscovered by E. Noether. We will establish the general Skolem-Noether Theorem later, in Sect. 4.9, when having appropriate tools at our disposal. Treating a special case at this early stage of development of the theory gives us an opportunity to demonstrate a simple, but efficient method of proof.

In unital rings (and algebras) there is a canonical way of constructing automorphisms; every invertible element gives rise to one:

Definition 1.26 An automorphism φ of a unital ring R is said to be an **inner automorphism** if there exists an invertible $a \in R$ such that $\varphi(x) = axa^{-1}, x \in R$.

An automorphism that is not inner is called **outer**. Here are a few examples of such automorphisms.

Example 1.27 The conjugation $z \mapsto \bar{z}$ is an automorphism of the $\mathbb{R}$-algebra $\mathbb{C}$. Of course it is outer for the identity map is obviously the only inner automorphism of a commutative ring. We also remark that it is an element of $\text{End}_{\mathbb{R}}(\mathbb{C})$, but not of $M(\mathbb{C})$. Therefore the assumption that A is central cannot be omitted in Lemma 1.25.

Example 1.28 For every $\alpha \in F \backslash \{0\}$, $f(\omega) \mapsto f(\omega + \alpha)$ is an outer automorphism of the algebra $F[\omega]$.

Example 1.29 Let S be a ring and let $R = S \times S$ be the direct product of two copies of S. Then $(s, t) \mapsto (t, s)$ is an outer automorphism of R.

In light of these examples one can appreciate the next theorem.

Theorem 1.30 (Skolem-Noether) *Every automorphism of a finite dimensional central simple algebra A is inner.*

Proof Let φ be an automorphism of A. Lemma 1.25 shows that $\varphi = \sum_{i=1}^{n} L_{a_i} R_{b_i}$ for some $a_i, b_i \in A$. We may assume that $a_1 \neq 0$ and the b_i's are linearly independent (cf. Remark 1.23). Since φ preserves multiplication, we have $L_{\varphi(x)}\varphi = \varphi L_x$ for every $x \in A$. Note that this yields

$$\sum_{i=1}^{n} L_{\varphi(x)a_i - a_i x} R_{b_i} = 0.$$

Lemma 1.24 in particular implies that

$$\varphi(x)a_1 - a_1 x = 0, \quad x \in A. \tag{1.11}$$

It remains to show that a_1 is invertible. As A is simple, there are $w_j, z_j \in A$ such that $\sum_{j=1}^{m} w_j a_1 z_j = 1$. In view of (1.11) this can be written as

$$\Big(\sum_{j=1}^{m} w_j \varphi(z_j)\Big) a_1 = 1, \quad \text{and} \quad a_1 \Big(\sum_{j=1}^{m} \varphi^{-1}(w_j) z_j\Big) = 1.$$

Having both a left and right inverse, a_1 is invertible.   □

Example 1.27 shows that the assumption that A is central is indispensable in Theorem 1.30.

Remark 1.31 (a) In light of the end of the proof of Theorem 1.30 it seems appropriate to mention that in a finite dimensional unital algebra A, a right inverse of an element is automatically also a left inverse; that is, $ab = 1$ implies $ba = 1$. This can be proved as follows. Take $a \in A$. By Lemma 1.5 there exists a nonzero polynomial $f(\omega) \in F[\omega]$ such that $f(a) = 0$. Assume that f has the smallest degree among all such polynomials. We can write $f(a) = 0$ as $ag(a) = \alpha$ where α is the negative of the constant term of f, and $g(\omega) \in F[\omega]$ is a nonzero polynomial with smaller degree than $f(\omega)$, so that $g(a) \neq 0$. If $\alpha \neq 0$, then a is invertible, $a^{-1} = \alpha^{-1}g(a)$. If $\alpha = 0$, then $ag(a) = g(a)a = 0$ and a cannot have neither a left nor right inverse.

(b) The discussion in (a) shows, in particular, that finite dimensional unital algebras enjoy a rather special property: every nonzero element is either invertible or a zero-divisor.

(c) A nonzero subalgebra A of a finite dimensional division algebra D is itself a division algebra. Indeed, take $0 \neq a \in A$; then $1 \in A$ for a is algebraic and invertible, and with reference to (a) we see that $a^{-1} = \alpha^{-1}g(a) \in A$.

The field of rational functions $F(\omega)$ is a division algebra over F, but its subalgebra $F[\omega]$ is not a division algebra. The finite dimensionality assumption in (c) is thus necessary. The next example shows that (a) and (b) do not always hold in infinite dimensional algebras.

Example 1.32 From calculus we know that differentiation and integration are inverse operations. But one needs to be a little careful to interpret this statement correctly. If by integration we mean the operator J given by $J(f)(x) = \int_0^x f(t)dt$, then by first integrating and then differentiating f we get f, but if we first differentiate and then integrate the constant function 1 we get 0.

Let us translate this phenomenon to an appropriate algebraic setting. Let D, J be elements in the algebra $\mathrm{End}_{\mathbb{R}}(\mathbb{R}[\omega])$ defined by $D(\omega^n) = n\omega^{n-1}$ (just as in Example 1.13) and $J(\omega^n) = \frac{\omega^{n+1}}{n+1}$, $n \geq 0$ (note that it is enough to specify the action of a linear operator on a basis). Then $DJ = I$, but $JD \neq I$. Thus, D is right invertible but not left invertible, and J is left invertible but not right invertible. From $D(I - JD) = 0$ and $TDJ = T$ for any $T \in \mathrm{End}_{\mathbb{R}}(\mathbb{R}[\omega])$ we infer that D is a left zero-divisor but not a right zero-divisor. Similarly, J is a right zero-divisor but not a left zero-divisor.

We have chosen this example because of its connection to calculus. Here is a slightly simpler example of two operators A and B that satisfy $AB = I$ and $BA \neq I$: $A(1) = 0, A(\omega^n) = \omega^{n-1}, n \geq 1$, and $B(\omega^n) = \omega^{n+1}, n \geq 0$.

1.7 Maximal Subfields

In this section we get back on track and consider division algebras again. Nevertheless, we will introduce the next notions in a more general setting.

If a subalgebra K of a unital algebra A is a field and $K \supseteq F$ (i.e., K contains the unity of A), then we call it a **subfield** of A. In this case we may consider A as a vector space over K. Indeed, one can regard the multiplication of elements from K with elements from A as a scalar multiplication $K \times A \to A$. The vector space axioms are readily fulfilled. We remark that this does not yet mean that A is also a K-algebra—this is true only if K is contained in the center $Z(A)$ of A (cf. Example 1.21).

Remark 1.33 If K is a subfield of a finite dimensional unital algebra A, then we have

$$[A : F] = [A : K][K : F].$$

This formula is well-known in the special case where A is an extension field of K. The same proof works in the general case. Indeed, if $\{k_i \mid i = 1, \dots, m\}$ is a basis of K over F, and $\{a_j \mid j = 1, \dots, n\}$ is a basis of A over K, then $\{k_i a_j \mid i = 1, \dots, m, j = 1, \dots, n\}$ is easily seen to be a basis of A over F.

Definition 1.34 A subfield that is not properly contained in a larger subfield of A is called a **maximal subfield** of A.

Example 1.35 Each of $\mathbb{R} \oplus \mathbb{R}i$, $\mathbb{R} \oplus \mathbb{R}j$, and $\mathbb{R} \oplus \mathbb{R}k$ is a maximal subfield of $\mathbb{H}$, 2-dimensional and isomorphic to $\mathbb{C}$. With some abuse of notation, we can therefore regard $\mathbb{H}$ as a complex vector space (but not as a complex algebra).

If K is a subfield of A, then $R_u \in \mathrm{End}_K(A)$ for every $u \in A$. Indeed, $R_u(kx) = kxu = kR_u(x)$ for all $k \in K, x \in A$. We also remark that given $f \in \mathrm{End}_K(A)$, its scalar multiple kf, where $k \in K$, can be interpreted as the operator $L_k f$. This is a trivial observation, but should be kept in mind when reading the next proof.

Theorem 1.36 *Let D be a finite dimensional central division F-algebra. If K is a maximal subfield of D, then* $[D : F] = [K : F]^2$.

Proof Pick a basis $\{u_1, \ldots, u_d\}$ of the F-algebra D. We claim that $\{R_{u_1}, \ldots, R_{u_d}\}$ is a basis of the K-algebra $\mathrm{End}_K(D)$. Lemma 1.24 implies that this set is linearly independent, so it suffices to show that it spans $\mathrm{End}_K(D)$. Let $f \in \mathrm{End}_K(D)$. In particular, $f \in \mathrm{End}_F(D)$, so it follows from Lemma 1.25 (and Remark 1.23) that $f = \sum_{i=1}^d L_{a_i} R_{u_i}$ for some $a_i \in D$. Since f is K-linear, i.e., $f L_k = L_k f$ holds for every $k \in K$, we have

$$\sum_{i=1}^d L_{a_i k - k a_i} R_{u_i} = 0.$$

Lemma 1.24 tells us that $a_i k - k a_i = 0$ for every i and every $k \in K$. This implies that $a_i \in K$, since otherwise the subalgebra generated by a_i and K would be a field larger than K (cf. Remark 1.31 (c)). Thus, f is a K-linear combination of the R_{u_i}'s, proving our claim.

Accordingly,

$$[D : F] = d = [\mathrm{End}_K(D) : K] = [D : K]^2.$$

On the other hand,

$$[D : F] = [D : K][K : F]$$

by Remark 1.33, and the result follows.   □

Corollary 1.37 *The dimension of a finite dimensional central division algebra is a perfect square.*

Proof A finite dimensional central division algebra certainly contains maximal subfields. Indeed, these are exactly the subfields of maximal dimensions. The desired conclusion therefore follows from Theorem 1.36.   □

The main application of Theorem 1.36 will be given in the next section.

1.8 Wedderburn's Theorem on Finite Division Rings

Finite fields exist and their structure is well-known. Do there exist finite division rings that are not commutative? The set of nonzero elements of such a ring would form a nonabelian finite group under multiplication. Understanding this group, which misses only one element in our ring, should be sufficient for understanding the ring. As finite groups are one of the most studied algebraic objects, it seems natural to involve group-theoretic techniques when addressing the above question.

We shall need the *class formula* from elementary group theory. Let us recall it and sketch its proof. Let G be a finite group with center $3(G)$. For every $a \in G$ we let

$$\mathfrak{C}(a) := \{g \in G \mid ag = ga\}.$$

This is a subgroup of G. The set $\{xax^{-1} \mid x \in G\}$ of all conjugates of a has the same cardinality as the set of all (left) cosets of $\mathfrak{C}(a)$ in G. Indeed, $xax^{-1} \mapsto x\mathfrak{C}(a)$ is a well-defined bijection. Conjugacy is an equivalence relation on G. Partitioning G into its disjoint equivalence classes (which are called conjugacy classes) and using Lagrange's theorem telling us that the number of cosets is $\frac{|G|}{|\mathfrak{C}(a)|}$, we obtain the desired formula

$$|G| = |3(G)| + \sum \frac{|G|}{|\mathfrak{C}(a)|}, \tag{1.12}$$

where the sum is taken over representatives of nontrivial conjugacy classes.

We are now in a position to prove that the answer to our question from the first paragraph is "no". This was established by J. H. M. Wedderburn in 1905.

Theorem 1.38 (Wedderburn) *A finite division ring is commutative.*

Proof Suppose this is not true. Then there exists a noncommutative finite division ring D of minimal cardinality; all its proper division subrings are therefore commutative. We may regard D as a central division algebra over its center F.

For every $a \in D$ we set

$$C(a) := \{x \in D \mid ax = xa\}.$$

Clearly $C(a)$ is a division subring, and hence a subfield of D if $a \notin F$. Since elements in any subfield that contains a commute with a, $C(a)$ is actually a maximal subfield. Let $q = |F|$. (As we know, q is a prime power, but we will not use this.) Note that an m-dimensional vector space over F has q^m elements. Accordingly, by Theorem 1.36 there exists $d \geq 2$ such that $|D| = q^{d^2}$ and $|C(a)| = q^d$ for every $a \in D \backslash F$. Now apply (1.12) to $G := D^* = D \backslash \{0\}$. Since $3(G) = F \backslash \{0\}$ and $\mathfrak{C}(a) = C(a) \backslash \{0\}$ it follows that there exists $s \in \mathbb{N}$ such that

$$q^{d^2} - 1 = q - 1 + s \frac{q^{d^2} - 1}{q^d - 1}.$$

Since $q^{d^2} - 1$ is a multiple of $\frac{q^{d^2}-1}{q^d-1}$, this leads to a contradiction that

$$\frac{q^{d^2} - 1}{q^d - 1} = \frac{(q^d)^d - 1}{q^d - 1} = 1 + q^d + \cdots + q^{d(d-2)} + q^{d(d-1)}$$

is a divisor of $q - 1$. $\qquad\qquad\qquad\qquad\qquad\qquad\qquad\qquad\qquad\qquad\qquad$ $\square$

The statement of this theorem is incredibly simple. It provides us an elementary connection between three fundamental concepts: invertibility, finiteness, and commutativity. Such simple, basic results often have equally simple, few lines proofs. But no such proof of Wedderburn's theorem is known (nor expected). There are, however, numerous different proofs, both elementary and advanced. Probably the most famous one is a short proof found by E. Witt in 1931. Although Witt's proof is a rather standard choice for a textbook (for example, it is included in Lam's book [Lam01]), we have demonstrated the proof by T. Nagahara and H. Tominaga from 1974. It is also quite short and, hopefully the reader will agree, beautiful. Besides, it is much closer to the spirit of this book than Witt's and many other proofs, especially because of the auxiliary results led to it, which will also be used later.

1.9 Further Examples of Division Algebras

So far we have met only one example of a noncommutative division algebra, namely $\mathbb{H}$. Moreover, we have proved that there are no noncommutative finite dimensional division algebras over algebraically closed fields or over finite fields, and that $\mathbb{H}$ is the only such example over the field of real numbers. The lack of examples has now become apparent. The aim of this section is to bring to light some relatively simple examples; there are other important examples which, however, are not so easily approachable by elementary means. Let us warn the reader that some verifications will be omitted in our discussion. We feel that without indicating additional examples this chapter would be a torso, but studying them in detail would make exposition a bit lengthy and tedious.

Example 1.39 In Remark 1.7 we have pointed out that one does not need to limit to $\mathbb{R}$ to get division algebras with the same multiplication table as $\mathbb{H}$. The idea now is to alter this table a little bit to get new examples of division algebras. Let F be a field with $\text{char}(F) \neq 2$, and pick $\lambda, \mu \in F\backslash\{0\}$. We define the **generalized quaternion algebra** $A = \left(\frac{\lambda,\mu}{F}\right)$ as the 4-dimensional unital algebra with basis $\{1, i, j, k\}$ and multiplication table

$$i^2 = \lambda, \quad j^2 = \mu, \quad k^2 = -\lambda\mu,$$

$$ij = -ji = k, \quad ki = -ik = -\lambda j, \quad jk = -kj = -\mu i.$$

Thus, $\mathbb{H} = \left(\frac{-1,-1}{\mathbb{R}}\right)$. It is easy to check that A is a central simple algebra; the proof is left as an exercise for the reader. We define the conjugate $\bar{a}$ of $a \in A$ in the same way as in $\mathbb{H}$. That is, if $a = \alpha_0 + v, \alpha_0 \in F, v = \alpha_1 i + \alpha_2 j + \alpha_3 k$, then $\bar{a} = \alpha_0 - v$. Noticing that

$$a\bar{a} = \bar{a}a = \alpha_0^2 - \lambda\alpha_1^2 - \mu\alpha_2^2 + \lambda\mu\alpha_3^2$$

we see, by the same argument as in Remark 1.7, that $A = \left(\frac{\lambda, \mu}{F}\right)$ is a division algebra if and only if for all $\alpha_i \in F$,

$$\alpha_0^2 - \lambda\alpha_1^2 - \mu\alpha_2^2 + \lambda\mu\alpha_3^2 = 0 \implies \alpha_0 = \alpha_1 = \alpha_2 = \alpha_3 = 0.$$

Now it depends on F whether this can be fulfilled for some choices of λ and μ. Of course it is never fulfilled if F is algebraically closed. If $F = \mathbb{R}$ then it is fulfilled if both λ and μ are < 0, but for any such choice of λ and μ we get an algebra isomorphic to $\mathbb{H}$. It turns out that over some fields (including $\mathbb{Q}$) we can get infinitely many pairwise nonisomorphic division algebras.

The downside of Example 1.39 is the limitation to dimension 4. By Corollary 1.37 this is the smallest possible dimension of a noncommutative central division algebra. Let us proceed to higher dimensions.

A class of rings is often closed under some constructions. Is it possible to form new division rings from a given division ring D? The usual suspects, direct products and matrix ring construction, obviously fail because of abundance of zero-divisors. What about the polynomial ring $D[\omega]$? It is even farther from a division ring, as polynomials of degree ≥ 1 will never be invertible in $D[\omega]$. Let us consider a larger **ring $D[[\omega]]$ of formal power series** over D. Its elements are written as formal sums $\sum_{n=0}^{\infty} a_n\omega^n$, $a_n \in D$. The main difference from polynomials is that we impose no restriction on the a_i's. The addition and multiplication are defined in the same way as for polynomials:

$$\sum_{n=0}^{\infty} a_n\omega^n + \sum_{n=0}^{\infty} b_n\omega^n := \sum_{n=0}^{\infty} (a_n + b_n)\omega^n,$$

$$\left(\sum_{n=0}^{\infty} a_n\omega^n\right)\left(\sum_{n=0}^{\infty} b_n\omega^n\right) := \sum_{n=0}^{\infty} c_n\omega^n, \text{ where } c_n = \sum_{i=0}^{n} a_i b_{n-i}.$$

This definition obviously makes sense if we replace D by an arbitrary ring R. Thus, for every ring R we can construct the **formal power series ring $R[[\omega]]$** which contains the polynomial ring $R[\omega]$ as a subring. However, let us return to the case where $R = D$.

Lemma 1.40 *Let D be a division ring. Then $\sum_{n=0}^{\infty} a_n\omega^n$ is invertible in $D[[\omega]]$ if and only if $a_0 \neq 0$.*

Proof The right invertibility of $\sum_{n=0}^{\infty} a_n\omega^n$ is equivalent to the existence of elements $b_i \in D, i \geq 0$, such that

$$a_0 b_0 = 1, \quad a_0 b_1 + a_1 b_0 = 0, \quad a_0 b_2 + a_1 b_1 + a_2 b_0 = 0, \text{ etc.}$$

If $a_0 \neq 0$, then these equations can be solved. Indeed, the first equation gives $b_0 = a_0^{-1}$, the second one gives $b_1 = -a_0^{-1} a_1 b_0$, etc. Thus in this case $\sum_{n=0}^{\infty} a_n\omega^n$ is right invertible. Similarly we see that it is left invertible, and hence invertible. Conversely, if $\sum_{n=0}^{\infty} a_n \omega^n$ is invertible, then $a_0 \neq 0$ for $a_0 b_0 = 1$ holds for some $b_0 \in R$.   □

Thus, although much closer than $D[\omega]$, $D[[\omega]]$ is still far from being a division ring. If only ω was invertible, then the invertibility of elements with zero constant term would not be a problem. Why not make it invertible?

Example 1.41 For an arbitrary ring R we define the **Laurent series ring** $R((\omega))$ as the set of all formal series $\sum_{n=-\infty}^{\infty} a_n \omega^n$ where only finitely many a_n's with $n < 0$ can be nonzero, and with operations defined in the obvious way by formally adding and multiplying series. Note that multiplication is possible since, by assumption, for every $f \neq 0$ in $R((\omega))$ there exists $m \in \mathbb{Z}$ such that $f = \sum_{n=m}^{\infty} a_n \omega^n$ and $a_m \neq 0$. One can check that $R((\omega))$ is a ring containing $R[[\omega]]$ as a subring. If $R = D$ is a division ring and f is a nonzero element written as above, then $f\omega^{-m}$ is invertible by Lemma 1.40, implying that f is invertible. Accordingly, $D((\omega))$ is a division ring.

If A is an algebra, then the ring $A((\omega))$ becomes an algebra by defining scalar multiplication in the self-explanatory way. Of course, if D is a division algebra, then $D((\omega))$ is also a division algebra. However, it is infinite dimensional. Therefore this example is not entirely satisfactory for us. Let us modify it a little bit.

In each of the rings $R[\omega]$, $R[[\omega]]$, and $R((\omega))$, ω commutes with elements in R and is therefore a central element. We will now change this as follows. Let σ be a ring endomorphism of R. Instead of $\omega a = a \omega$ we now require that $\omega a = \sigma(a)\omega$ for all $a \in R$. Accordingly,

$$(a\omega^i)(b\omega^j) = a\sigma^i(b)\omega^{i+j} \quad \text{for all } a, b \in R. \tag{1.13}$$

We can now consider the set of polynomials $\sum_{i=0}^{n} a_i \omega^i$ endowed with the usual addition and multiplication determined by (1.13). The resulting ring is called a **skew polynomial ring** and is denoted by $R[\omega; \sigma]$. Similarly we introduce the **skew power series ring** $R[[\omega; \sigma]]$. Assuming that σ is an automorphism we see that (1.13) makes sense also for the negative integers i and j, which enables us to introduce the **skew Laurent series ring** $R((\omega; \sigma))$.

Example 1.42 If D is a division algebra, then so is $D((\omega; \sigma))$. The proof is practically the same as for $D((\omega))$. Of course, $D((\omega; \sigma))$ is infinite dimensional over the base field F, but we can also view it as an algebra over its center Z. Let us add some assumptions. First of all, as the noncommutativity is guaranteed by (1.13), we assume that $D = K$ is a field. Further, assume that the automorphism σ has order n (i.e., n is the smallest natural number such that $\sigma^n = \text{id}_K$), and $F = \{x \in K \mid \sigma(x) = x\}$. Readers familiar with Galois Theory surely know concrete examples of such automorphisms. We leave it as an exercise to prove that the center Z of $K((\omega; \sigma))$ is $F((\omega^n))$, and that $[K((\omega; \sigma)) : Z] = n^2$ (or see [Lam01, pp. 217–218] for details). That is, $K((\omega; \sigma))$ is a central division algebra over Z of dimension n^2.

We recommend [Lam01, Pie82], and [Row08] for more examples of finite dimensional division algebras. One may say that these algebras do exist and are many, but it is not so easy to construct them. Also, it is not always easy to study them. Some rather basic questions about them are still open. Here is an example:

Problem 1.43 Suppose a division algebra D is algebraic and finitely generated (as an algebra). Is then D finite dimensional?

This question was posed for general (not necessarily division) algebras by A. G. Kurosh in the early 1940s. However, in 1964 E. S. Golod and I. R. Shafarevitch showed that in this generality the answer is negative. Thus it is necessary to impose some restrictions on the algebras in question. Problem 1.43 is often referred to as the "Kurosh problem for division algebras".

Exercises

1.1. Prove that the following statements are equivalent for a nonzero ring D:

 (i) D is a division ring.
 (ii) For all $a, b \in D$ with $a \neq 0$, the equations $ax = b$ and $ya = b$ have unique solutions in D.
 (iii) For all $a, b \in D$ with $a \neq 0$, the equation $ax = b$ has a solution in D.
 (iv) $D^2 \neq 0$ and D has no right ideals other than 0 and D.

 Remark: As mentioned in Sect. 1.2, nonassociative division algebras are defined through condition (ii). The equivalence of (i) and (ii) thus means that in associative algebras this definition coincides with the standard one. In nonassociative algebras, however, the condition of being a division algebra is independent of the condition that every nonzero element a has a right and left inverse. The latter only means that the equations $ax = b$ and $ya = b$ can be solved for $b = 1$, while the former does not even imply the existence of unity. For example, take an associative division algebra (possibly a field) D having an automorphism $\varphi \neq \mathrm{id}_D$. The vector space D endowed with the new product $x \cdot y := \varphi(xy)$ is then a nonassociative division algebra without unity.

1.2. Let x, y be elements in a division ring D such that $x \neq 0$, $y \neq 0$, and $x \neq y^{-1}$. Prove **Hua's identity** $xyx = x - \left(x^{-1} + (y^{-1} - x)^{-1}\right)^{-1}$.

1.3. Show that a division ring D satisfying $(xy)^2 = (yx)^2$ for all $x, y \in D$ is commutative.

 Hint: When considering an identity in a ring that involves arbitrary elements it is often useful to replace these elements by the sum (or product) of two elements.

1.4. Let D be a unital $\mathbb{R}$-algebra. Suppose there exists a linear map $x \mapsto \tilde{x}$ from D into D such that $x\tilde{x} = \tilde{x}x \in \mathbb{R}$ for every $x \in D$, and $x\tilde{x} \neq 0$ if $x \neq 0$. Show that D is isomorphic to $\mathbb{R}$, $\mathbb{C}$, or $\mathbb{H}$.

Hint: See the comment following Lemma 1.5, and the hint from the previous exercise.

1.5. Let $\alpha \in \mathbb{R}$. Find all solutions of the equation $x^2 = \alpha$ in $\mathbb{H}$.

1.6. Let R be a unital ring. Show that every ideal of $M_n(R)$ is of the form $M_n(I)$ for some ideal I of R.

1.7. Show that a (not necessarily unital) ring R is simple if and only if $M_n(R)$ is simple.

1.8. Show that the center of a simple ring is either 0 or a field. In particular, a commutative simple ring is thus necessarily a field.

1.9. Give an alternative proof of Proposition 1.6 by considering the eigenvalues of left multiplication maps.

1.10. Show that a ring R has a left unity (i.e., a left identity element with respect to multiplication) if and only if every map $f : R \to R$ satisfying $f(xy) = f(x)y$, $x, y \in R$, is of the form $f = L_a$ for some $a \in R$. Describe all such maps f in the case where $R = M_n(2\mathbb{Z})$.

1.11. Let A be a unital F-algebra with $[A : F] = n$. Show that $[M(A) : F] \geq n$, and find an example where $[M(A) : F] = n > 1$.

Remark: The other extreme where $[M(A) : F] = n^2$ is treated in Lemma 1.25.

1.12. Find an element in $M(\mathbb{H})$ that cannot be written as $L_a R_b$, $a, b \in \mathbb{H}$.

1.13. Let A be a finite dimensional central simple algebra. Suppose that a linear map $d : A \to A$ satisfies $d(xy) = d(x)y + xd(y)$, $x, y \in A$. Show that there exists $a \in A$ such that $d(x) = [a, x]$, $x \in A$.

Remark: In terminology introduced in Definition 2.41 and Example 2.43 below, this can be stated as that every derivation of A is inner.

1.14. Check that if φ is an automorphism of the algebra A, then $(a_{ij}) \mapsto (\varphi(a_{ij}))$ is an automorphism of the algebra $M_n(A)$. Use this to find an outer automorphism of the $\mathbb{R}$-algebra $M_n(\mathbb{C})$.

1.15. Show that a nonzero algebra endomorphism of a finite dimensional simple algebra is an automorphism.

1.16. Let A be a finite dimensional central simple algebra. Describe all algebra endomorphisms of $A \times A$.

1.17. Show that the only inner automorphism of the Weyl algebra $\mathscr{A}_1$ is $\mathrm{id}_{\mathscr{A}_1}$, and find an example of an outer automorphism of $\mathscr{A}_1$.

Hint: What restrictions on the action of an automorphism on D and L does $[D, L] = I$ bring?

1.18. A map $x \mapsto x^*$ from an F-algebra A into itself is called an **involution** if

$$(x + y)^* = x^* + y^*, \quad (xy)^* = y^* x^*, \quad (x^*)^* = x$$

for all $x, y \in A$. Immediate examples are the conjugation $h \mapsto \bar{h}$ on $\mathbb{H}$, the transposition $M \mapsto M^t$ on $M_n(F)$, and the conjugate transposition $M \mapsto M^*$ on $M_n(\mathbb{C})$. Given an involution $*$ on A, we set $S = \{x \in A \mid x^* = x\}$ and

$K = \{x \in A \mid x^* = -x\}$. Elements in S are called **symmetric**, and elements in K are called **skew-symmetric**. Show that S and K are additive subgroups of A, $xy + yx \in S$ for all $x, y \in S$, $xy - yx \in K$ for all $x, y \in K$, and $A = S \oplus K$ if $\mathrm{char}(F) \neq 2$. Find a linear involution on $M_2(F)$, $\mathrm{char}(F) \neq 2$, such that the space of symmetric elements S is 1-dimensional.

Hint: Involutions share some properties with the map $x \mapsto x^{-1}$ on the group of invertible elements, which is in the matrix case connected with the notion of an adjugate matrix.

1.19. A bijective linear map θ from an F-algebra A into itself is called an **antiautomorphism** if $\theta(xy) = \theta(y)\theta(x)$ for all $x, y \in A$ (linear involutions are thus antiautomorphisms). Describe all antiautomorphisms of $\mathbb{R}$-algebras $M_n(\mathbb{R})$ and $M_n(\mathbb{H})$.

Hint: What is the product of two antiautomorphisms?

1.20. Show that every maximal subfield K of $\mathbb{H}$ is of the form $K = h(\mathbb{R} \oplus \mathbb{R}i)h^{-1}$ for some $h \in \mathbb{H} \backslash \{0\}$.

Hint: This can be shown in many ways. To minimize computation, try to apply Theorem 1.30. More precisely, find an isomorphism from $\mathbb{R} \oplus \mathbb{R}i$ onto K, and then extend it to an automorphism of $\mathbb{H}$ (and thereby prove "by hand" a special case of Theorem 4.46 below).

1.21. Find a maximal subfield of the F-algebra $M_2(F)$ for $F = \mathbb{R}, \mathbb{C}$.

1.22. Prove that a finite dimensional division algebra D is a finite field if and only if there exists $n \in \mathbb{N}$ such that $x^n = x$ for all $x \in D$.

Remark: In Sect. 5.10 we will establish an incomparably deeper result. This exercise is meant just as a small illustration of the applicability of rudimentary tools that are at our disposal now.

1.23. Show that $\left(\frac{1,1}{F}\right) \cong M_2(F)$ for every field F with $\mathrm{char}(F) \neq 2$.

1.24. Let D be a division ring. Show that every nonzero ideal of $D[[\omega]]$ is generated by ω^n for some $n \geq 0$.

Chapter 2
Structure of Finite Dimensional Algebras

This chapter is centered around Wedderburn's structure theory which under rather mild and inevitable restrictions describes the form of a finite dimensional algebra. In some sense it reduces the problem of understanding finite dimensional algebras to the problem of understanding finite dimensional *division* algebras. The latter can be difficult, but at least for division algebras over some fields definitive answers were given in Chap. 1.

Just as in the previous one, in this chapter we also wish to follow the conceptual simplicity as our basic principle. Rather than building powerful general machineries and then derive classical theorems as their byproducts, we will give direct proofs that only illustrate more general phenomena. There will be enough opportunities for demonstrating more profound theories later. That said, we will usually say a bit more than needed for achieving our main goals, and make occasional digressions from the main themes. Partially this will be done in order to make the exposition less condensed and more illuminating, and partially because of future goals. Thus we shall give a survey of the very basic theory, but with one eye on more general topics.

2.1 Nilpotent Ideals

So far we have treated rings without proper nontrivial ideals. Such rings can be relatively easily handled, they are considered as "nice" from the structural point of view. We will now briefly consider rings that are in complete contrast; they contain nontrivial ideals with a particularly "bad" property.

Definition 2.1 An ideal I of a ring R is said to be a **nilpotent ideal** if there exists $n \in \mathbb{N}$ such that $I^n = 0$.

The condition $I^n = 0$ means that $u_1 u_2 \ldots u_n = 0$ for all $u_i \in I$. To avoid confusion, we introduce two related notions.

© Springer International Publishing Switzerland 2014
M. Brešar, *Introduction to Noncommutative Algebra*, Universitext,
DOI 10.1007/978-3-319-08693-4_2

Definition 2.2 An element a in a ring R is said to be a **nilpotent element** if there exists $n \in \mathbb{N}$ such that $a^n = 0$.

Definition 2.3 An ideal I of a ring R is said to be a **nil ideal** if all elements in I are nilpotent.

Obviously, a nilpotent ideal is nil. The converse is not true in general (see Example 2.7).

A trivial example of a nilpotent ideal is $I = 0$, but of course we are interested in nontrivial examples. An extreme case is when the ring R itself is nilpotent.

Example 2.4 Take any additive group R, and equip it with trivial product: $xy = 0$ for all $x, y \in R$. Then $R^2 = 0$.

Example 2.5 A nilpotent element lying in the center $Z(R)$ of the ring R clearly generates a nilpotent ideal. A simple example of such an element can be obtained by taking the direct product of the ring from Example 2.4 with any other ring.

The next example is more illuminating.

Example 2.6 Let $A = T_n(F)$ be the algebra of all *upper triangular $n \times n$ matrices* over a field F, i.e., matrices that have zeros below the main diagonal. Let N be the set of all *strictly upper triangular matrices*, i.e., matrices in A that have zeros also on the diagonal. Then N is an ideal of A such that $N^n = 0$ and $N^{n-1} \neq 0$. One can find a variety of other ideals of A that are contained in N, and are therefore nilpotent. For instance, the set I of all matrices whose $(1, n)$ entry is arbitrary and all other entries are 0 is an ideal of A satisfying $I^2 = 0$.

Example 2.7 Let A be the set of all infinite, $\mathbb{N} \times \mathbb{N}$ matrices over F that are upper triangular and have only finitely many nonzero entries. Note that A is an algebra under the standard matrix operations. Let I be the set of all matrices in A that are strictly upper triangular. Then I is a nil ideal of A which is not nilpotent.

Remark 2.8 **Nilpotent one-sided ideals** are defined in the same way as nilpotent ideals. If L is a nilpotent left ideal of R, then L is contained in a nilpotent (two-sided) ideal $L + LR$. Indeed, it is easy to see that $L^n = 0$ implies $(L + LR)^n = 0$. Similarly, every nilpotent right ideal is contained in a nilpotent ideal.

At this point we can mention a celebrated open problem. From Remark 2.8 we see that a ring with a nonzero nilpotent one-sided ideal also contains a nonzero nilpotent ideal. In 1930, G. Köthe conjectured that the same is true for nil ideals. The following question is thus known as **Köthe's problem**.

Problem 2.9 If a ring R has a nonzero nil one-sided ideal, does R also have a nonzero nil ideal?

In various special situations this is known to be true. But in general the problem is still open after all these years.

Let us return to nilpotent ideals.

Lemma 2.10 *The sum of two nilpotent ideals is a nilpotent ideal.*

Proof Let I, J be ideals such that $I^n = 0$ and $J^m = 0$. We claim that $(I+J)^{n+m-1} = 0$. That is, the product of $n + m - 1$ elements of the form $u + v$, $u \in I$, $v \in J$, is 0. Such a product can be written as a sum of products $w = w_1 w_2 \ldots w_{n+m-1}$ where each $w_i \in I \cup J$. If at least n of these $w_i's$ are in I, then $w = 0$ as $I^n = 0$. If the number of the w_i's belonging to I is smaller than n, then at least m of them lie in J, and hence $w = 0$ since $J^m = 0$. $\qquad\square$

By a **maximal nilpotent ideal** we mean a nilpotent ideal that is not properly contained in a larger nilpotent ideal.

Lemma 2.11 *If a ring R has a maximal nilpotent ideal N, then N contains all nilpotent ideals of R.*

Proof If I is another nilpotent ideal, then $I + N$ is again a nilpotent ideal by Lemma 2.10. Because of the maximality of N we must have $I + N = N$, and thus $I \subseteq N$. $\qquad\square$

Thus, a maximal nilpotent ideal, if it exists, is unique and is equal to the sum of all nilpotent ideals. However, not every ring has such an ideal. That is to say, the sum of all nilpotent ideals of a ring is not always nilpotent (although it is nil for each of its elements is contained in a nilpotent ideal by Lemma 2.10).

Example 2.12 Let A and I be the algebra and its nil ideal from Example 2.7. For every $k \in \mathbb{N}$, let I_k denote the set of all matrices in I with the property that their nonzero entries appear only in the first k rows. It is easy to check that I_k is a nilpotent ideal of A; in fact, $I_k^{k+1} = 0$. If A had a maximal nilpotent ideal N, then, by Lemma 2.11, N would contain each I_k, and hence also $\bigcup_{k=1}^{\infty} I_k = I$. However, I is not a nilpotent ideal.

Anyway, a finite dimensional algebra certainly does have a maximal nilpotent ideal. Indeed, it contains at least one nilpotent ideal (namely 0), and therefore the (nonempty) set of all nilpotent ideals contains an element of maximal dimension. This is obviously a maximal nilpotent ideal.

Definition 2.13 The maximal nilpotent ideal of a finite dimensional algebra A is called the **radical** of A.

Let us stress that this definition is adjusted to the finite dimensional context. There are several nonequivalent definitions of radicals of general rings, which agree with Definition 2.13 in the case of finite dimensional algebras.

Example 2.14 The radical of the algebra $A = T_n(F)$ from Example 2.6 is N, the set of all strictly upper triangular matrices. Indeed, N is a nilpotent ideal of A, and any ideal of A that properly contains N cannot be nilpotent since it necessarily contains a nonzero diagonal matrix.

Example 2.15 The radical of a finite dimensional simple algebra is 0. Namely, a simple algebra A cannot be nilpotent since $A^n = 0$ implies that A^{n-1} is a proper ideal of A, and hence it is 0.

From now on in this chapter we will be interested in rings that have no nonzero nilpotent ideals. At the end we will describe the structure of all finite dimensional algebras with this property.

2.2 Prime and Semiprime Rings

The goal of this section is to introduce two important classes of rings, the prime rings and the semiprime rings. Let us start, however, with another class with which the reader is presumably already familiar, at least in the commutative context.

Definition 2.16 A ring R is said to be a **domain** if for all $a, b \in R$, $ab = 0$ implies $a = 0$ or $b = 0$.

In other words, R is a domain if it has no left (or right) zero-divisors. Equivalently, R is a domain if it has the *cancellation property*: If $a \neq 0$, then each of $ab = ac$ and $ba = ca$ implies $b = c$.

Commutative domains are of utmost importance in algebra. Noncommutative domains also form a notable class of rings, but their role in noncommutative algebra is not entirely parallel to the role of commutative domains in commutative algebra. The most basic examples of commutative rings are domains, while even the matrix ring $M_n(F)$, which one might consider as the prototype of a noncommutative ring, is not a domain. The next lemma introduces a wider class of rings, which can be regarded as a noncommutative counterpart of the class of commutative domains.

Lemma 2.17 *Let R be a ring. The following conditions are equivalent:*

(i) *For all $a, b \in R$, $aRb = 0$ implies $a = 0$ or $b = 0$.*
(ii) *For all left ideals I and J of R, $IJ = 0$ implies $I = 0$ or $J = 0$.*
(iii) *For all right ideals I and J of R, $IJ = 0$ implies $I = 0$ or $J = 0$.*
(iv) *For all ideals I and J of R, $IJ = 0$ implies $I = 0$ or $J = 0$.*

Proof If I and J are left ideals satisfying $IJ = 0$, then $IRJ = 0$ since $RJ \subseteq J$. Hence we see that (i) implies (ii). Similarly, (i) implies (iii), and each of (ii) and (iii) trivially implies (iv).

Assume that (iv) holds and that $a, b \in R$ satisfy $aRb = 0$. The product of the ideals RaR and RbR is then 0. Hence one of them, say RaR, is 0. This implies that Ra and aR are two-sided ideals such that $Ra \cdot R = R \cdot aR = 0$. By (iv), $Ra = aR = 0$. But then $\mathbb{Z}a$ is an ideal of R satisfying $\mathbb{Z}a \cdot R = 0$, and so (iv) now yields $a = 0$. Thus, (iv) implies (i).   □

Definition 2.18 A ring R is said to be **prime** if it satisfies one (and hence all) of the conditions of Lemma 2.17.

Lemma 2.19 *A commutative ring is prime if and only if it is a domain.*

Proof It is enough to observe that $ab = 0$ implies $aRb = 0$ if R is commutative. $\square$

Remark 2.20 Let R be a prime ring with char$(R) \neq 0$. We first remark that if $0 \neq a \in R$ and $n \in \mathbb{N}$ are such that $na = 0$, then $aR(nb) = 0$ for every $b \in R$, and hence $nR = 0$. Secondly, if $nR = 0$ and $n = rs$ for some $r, s \in \mathbb{N}$, then $rR \cdot sR = 0$. Since rR and sR are ideals of R, we must have $rR = 0$ or $sR = 0$. This implies that char(R) is a prime number p. Therefore we can consider R as an algebra over $\mathbb{Z}_p$ in the natural way. That is, for $k \in \mathbb{Z}_p = \{0, 1, \ldots, p - 1\}$ and $x \in R$ we define kx as $x + \cdots + x$ (k times).

We proceed with a yet wider class of rings.

Lemma 2.21 *Let R be a ring. The following conditions are equivalent:*

(i) *For all $a \in R$, $aRa = 0$ implies $a = 0$.*
(ii) *For all left ideals I of R, $I^2 = 0$ implies $I = 0$.*
(iii) *For all right ideals I of R, $I^2 = 0$ implies $I = 0$.*
(iv) *For all ideals I of R, $I^2 = 0$ implies $I = 0$.*
(v) *R has no nonzero nilpotent ideals.*

Proof The proof that (i)–(iv) are equivalent is similar to the proof of Lemma 2.17, so we omit it. It is trivial that (v) implies (iv). Conversely, if (iv) holds and I is an ideal such that $I^n = 0$, then $(I^{n-1})^2 = 0$ and hence $I^{n-1} = 0$. Inductively, we get $I = 0$. $\square$

Definition 2.22 A ring R is said to be **semiprime** if it satisfies one (and hence all) of the conditions of Lemma 2.21.

We remark that as a special case of (i) we have that each of the conditions $aR = 0$ and $Ra = 0$ implies $a = 0$ if R is semiprime.

Let us record the obvious analogue of Lemma 2.19.

Lemma 2.23 *A commutative ring is semiprime if and only if it has no nonzero nilpotent elements.*

Remark 2.24 If A is an algebra and I, J are ring ideals of A such that $IJ = 0$, then the linear spans of I and J are algebra ideals of A whose product is 0. Hence it follows that a **prime algebra** can be equivalently defined as an algebra which is prime as a ring, or as an algebra in which the product of any two of its nonzero algebra ideals is nonzero. A similar remark holds for **semiprime algebras** (as well as for simple algebras, cf. Remark 1.11).

Remark 2.25 Let I be a nonzero ideal of a prime ring R. If $a, b \in R$ are such that $aIb = 0$, then $aRuRb = 0$ for every $u \in I$. Using (i) from Lemma 2.17 twice it follows that $a = 0$ or $b = 0$. This in particular shows that an ideal of a prime ring is again a prime ring. Similarly, an ideal of a semiprime ring is a semiprime ring.

Let us return to the situation considered in Lemma 2.11. The following can be added.

Lemma 2.26 *If N is a maximal nilpotent ideal of a ring R, then the factor ring R/N is semiprime.*

Proof Let K be an ideal of R/N such that $K^2 = 0$. Then $J = \{x \in R \mid x + N \in K\}$ is an ideal of R such that $J^2 \subseteq N$. As N is nilpotent, it follows that J is nilpotent as well. Lemma 2.11 therefore implies that $J \subseteq N$, and hence $K = 0$.   □

Example 2.27 Let A and N be as in Example 2.14. Then A/N is isomorphic to F^n, the direct product of n copies of F. Indeed, the map $(a_{ij}) \mapsto (a_{11}, a_{22}, \ldots, a_{nn})$ from A to F^n is easily seen to be a surjective algebra homomorphism with kernel N.

The following relations between the classes of rings introduced so far are obvious from definitions:

$$\text{division ring} \implies \text{simple and domain},$$
$$\text{simple} \implies \text{prime},$$
$$\text{domain} \implies \text{prime},$$
$$\text{prime} \implies \text{semiprime}.$$

None of these implications can be reversed.

Example 2.28 We know that the Weyl algebra $\mathscr{A}_1$ is simple. Let us show that it is also a domain, but not a division ring. We use the same notation as in Example 1.13. Take $S, T \in \mathscr{L}$ and $r, s \geq 0$. Applying (1.8), it follows that

$$(SD^r)(TD^s) = STD^{r+s} + S[D^r, T]D^s = STD^{r+s} + \sum_{i=s}^{r+s-1} S_i D^i$$

for some $S_i \in \mathscr{L}$. Accordingly, for all $S_r, T_s \in \mathscr{L}$ we have

$$\left(\sum_{r=0}^{m} S_r D^r\right)\left(\sum_{s=0}^{n} T_s D^s\right) = S_m T_n D^{m+n} + \sum_{j=0}^{m+n-1} S_j' D^j$$

where $S_j' \in \mathscr{L}$. Since $\mathscr{L}$ is obviously a domain (isomorphic to $F[\omega]$), it follows from (b) in Example 1.13 that $\mathscr{A}_1$ is a domain, and also that the only invertible elements in $\mathscr{A}_1$ are the nonzero elements from F.

Example 2.29 The ring $\mathbb{Z}$ is a domain that is not a simple ring.

Example 2.30 The matrix ring $M_n(F)$, $n \geq 2$, is simple, but not a domain.

Example 2.31 It is easy to check that the ring $M_n(\mathbb{Z})$, $n \geq 2$, is prime. But it is neither a domain nor a simple ring. The obvious, and in fact the only examples of its ideals are $M_n(k\mathbb{Z})$, $k \geq 0$.

Example 2.32 Every commutative ring that has zero-divisors but does not have nonzero nilpotent elements, is semiprime but not prime. A concrete example is the ring of continuous functions $C[a, b]$.

The next example points out an important difference between the classes of prime and semiprime rings, and also indicates at least a slight analogy between prime rings and prime numbers.

Example 2.33 Let R_1 and R_2 be nonzero rings. Then their direct product $R = R_1 \times R_2$ is not a prime ring, as $R_1 \times 0$ and $0 \times R_2$ are nonzero ideals of R whose product is 0. On the other hand, if both R_1 and R_2 are semiprime, then R is also semiprime.

We shall soon meet an important and non-obvious example of a finite dimensional semiprime algebra, related to the notion of a group. Before arriving at this specific topic, we will consider, in the next two sections, some notions that are of general importance in ring theory.

2.3 Unitization

Let A be an F-algebra. Then the set $F \times A$ becomes an F-algebra, which we denote by $A^\sharp$, if we define addition, scalar multiplication and product as follows:

$$(\lambda, x) + (\mu, y) := (\lambda + \mu, x + y),$$
$$\mu(\lambda, x) := (\mu\lambda, \mu x),$$
$$(\lambda, x)(\mu, y) := (\lambda\mu, \mu x + \lambda y + xy).$$

We consider A as a subalgebra of $A^\sharp$ via the embedding $x \mapsto (0, x)$. Note that A is actually an ideal of $A^\sharp$. A crucial observation for us is that $A^\sharp$ is a unital algebra. Indeed, $(1, 0)$ is its unity.

Definition 2.34 The algebra $A^\sharp$ is called the **unitization** of A.

Remark 2.35 Replacing the role of F by $\mathbb{Z}$ one defines the unitization $R^\sharp$ of a ring R in exactly the same way (just ignore the scalar multiplication). Alternatively, one can use $\mathbb{Z}_n$ instead of $\mathbb{Z}$ if char$(R) = n > 0$.

This construction is intended primarily for algebras (and rings) without unity. The idea behind it is to be able to reduce some problems for general algebras to unital algebras; this does not always work, but sometimes it does. In principle one can construct $A^\sharp$ even when A is unital. This may seem somewhat artificial at first glance, especially since the unity of A is then different from the unity of $A^\sharp$. However, constructions of new rings and algebras from the old ones sometimes turn out to be unexpectedly useful, say when searching for counterexamples.

The unitization of A does not preserve all properties of A. For instance, the simplicity is definitely not preservered since A is an ideal of $A^\sharp$. If A is a nonzero prime unital algebra, then $A^\sharp$ is not prime since $I = \{(\lambda, -\lambda) \mid \lambda \in F\}$ is an ideal of $A^\sharp$ such that $IA = AI = 0$. The non-unital case is different.

Lemma 2.36 *If A is a prime algebra without unity, then $A^\sharp$ is also prime.*

Proof Let $(\lambda, a), (\mu, b) \in A^\sharp$ satisfy $(\lambda, a)A^\sharp(\mu, b) = 0$. Then $(\lambda, a)(1, 0)(\mu, b) = 0$, and hence $\lambda = 0$ or $\mu = 0$. Let us consider the case where $\lambda = 0$; the case where $\mu = 0$ can be treated similarly. We may assume that $a \neq 0$. From $(0, a)(0, x)(\mu, b) = 0$ we infer that $\mu ax + axb = 0$ for every $x \in A$. We may now also assume that $\mu \neq 0$, since otherwise $aAb = 0$ and hence $b = 0$, as desired. Setting $e := -\mu^{-1}b$ we thus have $ax = axe$ for every $x \in A$. Accordingly, $a(xy)e = axy = (axe)y$ for all $x, y \in A$. We can rewrite this as

$$ax(y - ye) = 0 = ax(y - ey).$$

Since A is prime it follows that $y = ye$ and $y = ey$ for every $y \in A$. This contradicts the assumption that A is not unital. $\square$

2.4 The Regular Representation

Let us begin with a question. Given a unital algebra A, is it possible that

$$[a, b] = 1 \tag{2.1}$$

for some $a, b \in A$? It is not our intention to discuss the reasons for the relevance of this equation. This problem was primarily chosen in order to illustrate the usefulness of the concept that we are about to introduce.

The reader's answer to our question might be "why not?", and it is difficult to argue against it. After all, we saw that (2.1) appears in the Weyl algebra $\mathscr{A}_1$; see (1.4). This algebra is infinite dimensional. Let us modify our question: *Can* (2.1) *occur in a finite dimensional algebra*? The answer is closer as it may seem. A shortcut to it is based on left multiplication maps $L_a : x \mapsto ax$. Recalling the formulas $L_{\lambda a + \mu b} = \lambda L_a + \mu L_b$ and $L_{ab} = L_a L_b$ we see that $a \mapsto L_a$ is an algebra homomorphism.

Definition 2.37 The homomorphism $a \mapsto L_a$ from A into $\mathrm{End}_F(A)$ is called the **regular representation** of A.

The regular representation is injective, unless $aA = 0$ for some nonzero $a \in A$. The latter condition is quite special, in particular it cannot happen if A is unital. Thus, under a mild assumption $a \mapsto L_a$ is an embedding of A in $\mathrm{End}_F(A)$, which makes it possible for one to identify an element $a \in A$ with $L_a \in \mathrm{End}_F(A)$. When trying to deal with an element a from an abstract algebra one might feel barehanded at the start. The advantage of considering L_a is that it is a linear operator and therefore we can rely on linear algebra methods.

The answer to our question is now within reach, but let us first record the following lemma giving a firm basis for using the approach just indicated. The reader will notice the analogy with Cayley's theorem from group theory.

Proposition 2.38 *Every F-algebra A can be embedded into the algebra $\mathrm{End}_F(V)$ for some vector space V. If A is finite dimensional, then V can be chosen to be finite dimensional, and so in this case A can be embedded into $M_n(F)$ for some $n \in \mathbb{N}$.*

Proof If A is unital, then we can take $V = A$ and apply the regular representation. If it is not unital, then we can embed A into its unitization $A^{\sharp}$, and accordingly take $V = A^{\sharp}$. In any case V is finite dimensional if A is; if V is n-dimensional, then $\mathrm{End}_F(V) \cong M_n(F)$. $\qquad\square$

This proposition enables one to consider elements from a finite dimensional algebra as matrices. The proof was easy, and this observation is often useless. But occasionally it can be very helpful. Let us return to our question. Thus, assume that a finite dimensional algebra contains elements a, b satisfying (2.1). Then, by Proposition 2.38, there exist matrices $A, B \in M_n(F)$ such that $[A, B] = I$ (namely, the regular representation maps 1 into I). Now, the trace of the matrix $[A, B] = AB - BA$ is 0, while the trace of the matrix I is n. If $\mathrm{char}(F) = 0$, then this is impossible, and hence (2.1) cannot occur. The same argument yields the following sharper assertion.

Proposition 2.39 *Let A be a nonzero finite dimensional unital algebra over a field F with $\mathrm{char}(F) = 0$. Then 1 cannot be written as a sum of commutators in A.*

Example 2.40 Let us show that Proposition 2.39 does not hold if F has prime characteristic p. Define $D, L \in \mathrm{End}_F(F[\omega])$ just as in Example 1.13. Let I be the ideal of $F[\omega]$ generated by ω^p, let $V = F[\omega]/I$, and let $\phi : F[\omega] \to V$ be the canonical homomorphism. Note that V is a p-dimensional vector space. Since $D(\omega^p) = 0$ it follows that $D(I) \subseteq I$. We have $L(I) \subseteq I$ for trivial reasons. These inclusions show that the maps $d, \ell : V \to V$ given by $d\phi = \phi D$ and $\ell\phi = \phi L$ are well-defined. Clearly $d, \ell \in \mathrm{End}_F(V) \cong M_p(F)$, and we have $[d, \ell] = 1$. The reader may wish to write down the matrix representations of d and ℓ to obtain a more explicit interpretation of this example.

One might wonder whether there is really no other way to handle (2.1) than reducing the problem to matrices. This does seem to be the most natural and simple way, but there are others. Let us allow ourselves a digression, and prove a lovely result by N. Jacobson from 1935, dealing with the more general condition $[[a, b], a] = 0$. This can easily occur in finite dimensional algebras even when a and b do not commute, regardless of $\mathrm{char}(F)$. Say, the standard matrix units E_{11}, E_{12} satisfy $[E_{12}, E_{11}] = -E_{12}$, and hence $[[E_{12}, E_{11}], E_{12}] = 0$. The most convenient way to consider this condition is through the following concept.

Definition 2.41 Let A be an algebra (resp. ring). A linear (resp. additive) map d : $A \to A$ is called a **derivation** if $d(xy) = d(x)y + xd(y)$ for all $x, y \in A$.

It is easy to guess where this name comes from.

Example 2.42 The differential operator D from Example 1.13 is a derivation.

The basic example in noncommutative rings is of different nature.

Example 2.43 For every a in an algebra (or ring) A, the map $d : A \to A$ given by $d(x) = [a, x]$ is a derivation. Such a derivation is said to be an **inner derivation**.

Proposition 2.44 (Jacobson) *Let A be a finite dimensional algebra over a field F with* char$(F) = 0$. *If $a, b \in A$ satisfy $[[a, b], a] = 0$, then $[a, b]$ is a nilpotent element.*

Proof Let d be the inner derivation $d(x) = [a, x]$. Our condition can be written as $d^2(b) = 0$. A simple induction argument shows that

$$d(d(b)^k) = 0 \quad \text{for every } k \in \mathbb{N}. \tag{2.2}$$

We claim that

$$d^n(b^n) = n!d(b)^n \quad \text{for every } n \in \mathbb{N}. \tag{2.3}$$

This is trivial for $n = 1$, so let $n > 1$ and assume that (2.3) holds for $n - 1$. From the definition of a derivation one easily infers that d^n satisfies the Leibniz rule that we know from calculus. Accordingly,

$$d^n(b^n) = d^n(bb^{n-1}) = \sum_{i=0}^{n} \binom{n}{i} d^i(b)d^{n-i}(b^{n-1})$$
$$= bd(d^{n-1}(b^{n-1})) + nd(b)d^{n-1}(b^{n-1}).$$

Now use the induction assumption together with (2.2) and (2.3) follows.

As an immediate consequence of (2.2) and (2.3) we get $d^m(b^k) = 0$ whenever $k < m$. By Lemma 1.5 there exists $m \in \mathbb{N}$ such that b^m is a linear combination of b^k with $k < m$. Hence $d^m(b^m) = 0$, and so $m!d(b)^m = 0$ by (2.3); that is, $[a, b]^m = 0$. $\qquad\square$

The proof just given was actually discovered by D. Kleinecke, but it is similar to Jacobson's original proof.

Remark 2.45 From the proof we see that a slightly more general result than stated is true: If A is as in Proposition 2.44, d is a derivation of A, and $d^2(b) = 0$ for some $b \in A$, then $d(b)$ is nilpotent. Note that we can reword Proposition 2.44 as follows: If δ is an inner derivation of A (induced by b), and $a \in A$ is such that $[\delta(a), a] = 0$, then $\delta(a)$ is nilpotent. Using the regular representation it can be easily shown that this is also true for every, not necessarily inner, derivation δ. Indeed, $[\delta(a), a] = 0$ implies $[L_{\delta(a)}, L_a] = 0$. The condition that δ is a derivation can be expressed as $L_{\delta(a)} = [\delta, L_a]$. Therefore we have $[[\delta, L_a], L_a] = 0$. Proposition 2.44 applied to the (also finite dimensional) algebra $\mathrm{End}_F(A)$ tells us that $L_{\delta(a)} = [\delta, L_a]$ is nilpotent. Hence $\delta(a)$ is nilpotent too.

A more sophisticated application of the regular representation will be given in the next section.

2.5 Group Algebras

Group theory and ring theory are two branches of algebra. In a first course in abstract algebra one might get an impression that, although sometimes similar, they are basically unconnected. But there are many links between groups and rings. The concept that we are about to introduce gives rise to striking interactions.

Assume temporarily that G is an arbitrary set. If F is, as always, a field, then we can form the vector space over F whose basis is G. Its elements are formal sums $\sum_{g\in G} \lambda_g g$ where $\lambda_g \in F$ and all but finitely many λ_g are zero. The definitions of addition and scalar multiplication are self-explanatory. Assume now that G is a group. Then this vector space becomes an algebra if we define multiplication by simply extending the group multiplication on G to the whole space; taking into account the algebra axioms this can obviously be done in a unique way. Thus,

$$\left(\sum_{g\in G}\lambda_g g\right)\left(\sum_{h\in G}\mu_h h\right) = \sum_{k\in G} \nu_k k, \quad \text{where} \quad \nu_k = \sum_{gh=k}\lambda_g\mu_h.$$

We denote this algebra by $F[G]$.

Definition 2.46 The algebra $F[G]$ is called the **group algebra** of G over F.

It is actually enough to assume that G is merely a semigroup to construct $F[G]$. In this case we call $F[G]$ a **semigroup algebra**. Similarly, we can speak about a **monoid algebra** if G is a monoid. Furthermore, one can replace the role of the field F by any ring R, and then define addition and multiplication in formally the same way. Everything still makes sense, but of course the resulting object is a ring rather than an algebra. It is called a **(semi)group ring** and is denoted by $R[G]$. For example, considering $\mathbb{N}_0 := \mathbb{N} \cup \{0\}$ as a monoid under addition, we readily see that $R[\mathbb{N}_0]$ is isomorphic to the polynomial ring $R[\omega]$.

It is easy to find group algebras that are domains. Say, $F[\mathbb{Z}]$ is such an example. However, if a group G has an element $g \neq 1$ of finite order, then $F[G]$ is not a domain. Indeed, $g^n = 1$ implies $(1 - g)(1 + g + \cdots + g^{n-1}) = 0$. The following problem has been open for a long time.

Problem 2.47 If a group G has no elements different from 1 of finite order, is then $F[G]$ a domain?

If G is a nontrivial group, then the set of all $\sum_{g \in G} \lambda_g g$ such that $\sum_{g \in G} \lambda_g = 0$ is a proper nonzero ideal of $F[G]$. It is called the **augmentation ideal** of $F[G]$. The algebra $F[G]$ therefore is not simple. The main issue of this section is the question whether $F[G]$ is semiprime. We begin with the simplest example.

Example 2.48 Let $G = \mathbb{Z}_2$. It is more convenient to use the multiplicative notation, so we write $G = \{1, g\}$ where $g^2 = 1$. The group algebra $F[G]$ thus consists of elements of the form $\lambda 1 + \mu g$, $\lambda, \mu \in F$, which are multiplied according to

$$(\lambda 1 + \mu g)(\lambda' 1 + \mu' g) = (\lambda \lambda' + \mu \mu')1 + (\lambda \mu' + \mu \lambda')g.$$

It is easy to verify that $\lambda 1 + \mu g \mapsto (\lambda + \mu, \lambda - \mu)$ is a homomorphism from $F[G]$ into $F \times F$. If $\mathrm{char}(F) \neq 2$, then it is bijective; thus, $F[G] \cong F \times F$ holds in this case. In particular, $F[G]$ is semiprime. This is no longer true if $\mathrm{char}(F) = 2$ since $F(1+g)$ is then a nilpotent ideal of $F[G]$. Note that, in this case, $F[G]$ is isomorphic to the subalgebra of $M_2(F)$ consisting of matrices of the form $\begin{bmatrix} \alpha & \beta \\ 0 & \alpha \end{bmatrix}$. An isomorphism is given by $\lambda 1 + \mu g \mapsto \begin{bmatrix} \lambda + \mu & \mu \\ 0 & \lambda + \mu \end{bmatrix}$.

Thus, the group algebra $F[\mathbb{Z}_2]$ is semiprime if and only if $\mathrm{char}(F) \neq 2$. We will now generalize this observation to group algebras over arbitrary finite groups. First we recall some facts from linear algebra. Let V be a finite dimensional vector space, and let $T : V \to V$ be a linear map. Choosing a basis in V, we can represent T as a matrix relative to this basis. Different bases give rise to similar matrices. Therefore, no matter which basis we choose, the trace of a matrix representing T is always the same. We can therefore define $\mathrm{tr}(T)$, the **trace of** T, as the trace of any matrix representation of T. If T is nilpotent, then $\mathrm{tr}(T) = 0$. This can be deduced, for example, from the Jordan normal form of a matrix representation of T, as 0 is clearly the only eigenvalue of T. Note that here we have to use the fact that every field F has an algebraic closure $\overline{F}$, and then consider the matrix representation of T as a matrix in $M_n(\overline{F})$.

The following theorem was proved in 1898 by H. Maschke.

Theorem 2.49 (Maschke) *Let G be a finite group. Then the group algebra $F[G]$ is semiprime if and only if either $\mathrm{char}(F) = 0$ or $\mathrm{char}(F)$ is a prime number p that does not divide $|G|$.*

Proof Since $F[G]$ is a finite dimensional vector space over F, we can define a linear functional $\rho : F[G] \to F$ by

$$\rho(a) := \mathrm{tr}(L_a).$$

Let $n = |G|$ and denote the elements in G by $g_1, g_2, \ldots, g_n$ where $g_1 = 1$. Obviously, $\rho(g_1) = n$. If $i \geq 2$, then

$$L_{g_i}(g_j) = g_i g_j \in G \setminus \{g_j\}$$

for every j. The matrix representation of L_{g_i} with respect to the basis $\{g_1, g_2, \ldots, g_n\}$ has therefore zeros on the diagonal. Consequently, $\rho(g_i) = 0$. This is a crucial observation upon which our proof is based.

Suppose now that $F[G]$ has a nonzero nilpotent ideal I. We want to show that then F has finite characteristic p which divides $n = |G|$. Pick a nonzero $a \in I$, and write $a = \sum_{i=1}^{n} \lambda_i g_i$. Without loss of generality we may assume that $\lambda_1 \neq 0$. Indeed, otherwise we choose i such that $\lambda_i \neq 0$ and replace a by $g_i^{-1}a$, which is of course also an element from I. We have

$$\rho(a) = \lambda_1 \rho(g_1) + \lambda_2 \rho(g_2) + \cdots + \lambda_n \rho(g_n) = n\lambda_1.$$

As an element of a nilpotent ideal, a is a nilpotent element. Hence L_a is a nilpotent linear map, and so $\rho(a) = 0$. That is, $n\lambda_1 = 0$. Since $\lambda_1 \neq 0$, this is possible only when $p = \mathrm{char}(F)$ divides n.

Conversely, assume that $p = \mathrm{char}(F)$ divides $|G|$. Set $r = \sum_{i=1}^{n} g_i$. Since $rg_j = g_j r = r$ for every j, we see that the 1-dimensional space Fr is an ideal of $F[G]$. As $r^2 = |G|r = 0$, Fr is a nilpotent ideal. $\square$

The real meaning of Theorem 2.49 will become clearer later, after describing the structure of finite dimensional semiprime algebras. This description is, in fact, our central goal in the rest of the chapter. To this end we need several simple auxiliary results, which are all of independent interest.

2.6 Matrix Units

We are already familiar with *standard* matrix units E_{ij}. Let us introduce their abstract generalization.

Definition 2.50 Let R be a unital ring and let $n \in \mathbb{N}$. A set $\{e_{ij} \in R \mid 1 \leq i, j \leq n\}$ is called a **set of $n \times n$ matrix units** if

$$e_{11} + e_{22} + \cdots + e_{nn} = 1$$

and

$$e_{ij}e_{kl} = \delta_{jk}e_{il}$$

for all $1 \leq i, j, k, l \leq n$. Here, δ_{jk} stands for the "Kronecker delta": $\delta_{jk} = \begin{cases} 1 \text{ if } j = k \\ 0 \text{ if } j \neq k \end{cases}$.

Standard matrix units in $R = M_n(S)$, where S is an arbitrary unital ring, of course provide a basic example. This is not the only set of matrix units in R, not even when $S = F$ is a field. For example, if e_{ij} are matrix units and p is an invertible element, then $f_{ij} := p^{-1}e_{ij}p$ are also matrix units.

The matrix units e_{ij} with $i \neq j$ are substantially different from the matrix units e_{ii}. In particular, the e_{ij}'s satisfy $e_{ij}^2 = 0$ and are thus nilpotent elements, while the e_{ii}'s satisfy $e_{ii}^2 = e_{ii}$.

Definition 2.51 An element e in a ring R is said to be an **idempotent** if $e^2 = e$. Idempotents e and f are called **orthogonal** if $ef = fe = 0$.

Thus, the e_{ii}'s are pairwise orthogonal idempotents whose sum is 1. Every e_{ii} gives rise to the subring $e_{ii}Re_{ii} = \{e_{ii}ae_{ii} \mid a \in R\}$ of R, and all of these rings are isomorphic:

$$e_{ii}Re_{ii} \cong e_{jj}Re_{jj}.$$

Indeed,

$$e_{ii}ae_{ii} \mapsto e_{ji}(e_{ii}ae_{ii})e_{ij} = e_{jj}(e_{ji}ae_{ij})e_{jj}$$

is an isomorphism. Checking this is straightforward and left as an exercise. We just wanted to indicate why one often considers only $e_{11}Re_{11}$. Another instructive little exercise is to show that the ideal generated by each matrix unit e_{ij} is the whole ring.

We have defined matrix units in an arbitrary unital ring. However, we will now show that a ring with matrix units is in fact a full matrix ring.

Lemma 2.52 *If a unital ring R contains a set of $n \times n$ matrix units e_{ij}, then $R \cong M_n(S)$ where $S = e_{11}Re_{11}$.*

Proof Define $\varphi : R \to M_n(e_{11}Re_{11})$ by

$$\varphi(a) := (a_{ij}), \quad \text{where } a_{ij} = e_{1i}ae_{j1}.$$

Note that $a_{ij} = e_{11}a_{ij}e_{11}$ and so a_{ij} indeed lies in $e_{11}Re_{11}$. The additivity of φ is clear. The (i,j) entry of $\varphi(a)\varphi(b)$ is equal to

$$\sum_{k=1}^{n} e_{1i}ae_{k1}e_{1k}be_{j1} = e_{1i}a\Big(\sum_{k=1}^{n} e_{kk}\Big)be_{j1} = e_{1i}abe_{j1},$$

which is the (i,j) entry of $\varphi(ab)$. Thus, $\varphi(ab) = \varphi(a)\varphi(b)$. If $a_{ij} = 0$ for all i,j, then $e_{ii}ae_{jj} = e_{i1}a_{ij}e_{1j} = 0$, and so $a = 0$ since the sum of the e_{ii}'s is 1. Thus φ is injective. Finally, observe that $\varphi(e_{k1}ae_{1l})$ is the matrix whose (k, l) entry is $e_{11}ae_{11}$ and all other entries are 0, from which the surjectivity of φ follows. $\square$

Remark 2.53 If R is an algebra, then the isomorphism from Lemma 2.52 is an algebra isomorphism.

2.7 Idempotents

Let us take a brief digression and say a few general remarks about idempotents. In what follows e will denote an arbitrary idempotent in a ring R.

As we have seen in the preceding section, the subring eRe of R may be of relevance. We call eRe the **corner ring** corresponding to e. This term obviously arises from the matrix ring example (what is eRe if $R = M_n(S)$ and $e = E_{11} + \cdots + E_{kk}$?). The corner ring eRe is unital, even if R is not; its unity is e. We shall soon see that eRe is only one out of four subrings of R that are naturally attached to e.

For simplicity of exposition we assume, until further notice, that R is a unital ring and that e is a **nontrivial idempotent**, i.e., an idempotent different from 0 and 1. Then

$$f := 1 - e$$

is also a nontrivial idempotent, e and f are orthogonal, and their sum is 1. A model for such a pair of idempotents, and a motivating example for what we are about to say, are matrix units e_{11} and e_{22} in a 2×2 matrix ring.

Suppose $ex_1e + ex_2f + fx_3e + fx_4f = 0$ for some $x_i \in R$. Multiplying from the left and right by e we obtain $ex_1e = 0$. Similarly we see that all other terms are 0. On the other hand, since every $x \in R$ can be written as

$$x = exe + exf + fxe + fxf$$

it follows that

$$R = eRe \oplus eRf \oplus fRe \oplus fRf. \tag{2.4}$$

Here we have in mind the additive group direct sum. We call (2.4) the **Peirce decomposition** of R (with respect to e). Set

$$R_{11} := eRe, \quad R_{12} := eRf, \quad R_{21} := fRe, \quad R_{22} := fRf.$$

Thus R_{11} and R_{22} are corner rings corresponding to e and f, respectively, while R_{12} and R_{21} are subrings with zero multiplication. Furthermore, we have

$$R_{ij}R_{kl} \subseteq \delta_{jk}R_{il}$$

for all $1 \le i, j, k, l \le 2$, which resembles calculations with matrix units. The point here is that when having a nontrivial idempotent e in our ring, we can, with the help of its indispensable companion $f = 1 - e$, mimic 2×2 matrices. Sometimes such an approach is really efficient. We did not say, however, that a ring R with a nontrivial idempotent e is necessarily isomorphic to a 2×2 matrix ring (over some ring). This does hold true under the additional assumption that the equations

$$exfye = e \quad \text{and} \quad fyexf = f$$

are solvable in R. Namely, then we can introduce the elements

$$e_{11} := e, \quad e_{12} := exf, \quad e_{21} := fye, \quad e_{22} := f,$$

which readily form a set of 2×2 matrix units of R, yielding $R \cong M_2(eRe)$ by Lemma 2.52.

There are many rings without nontrivial idempotents, for example domains, and hence in particular division rings. As indicated above, the existence of a sole nontrivial idempotent can already have an impact on the handling of a ring. But actually the existence of one idempotent implies the existence of "many", at least when the summands eRf and fRe from the Peirce decomposition are nonzero. Namely, if e is an idempotent, then so are $e + exf$ and $e + fxe$ for every $x \in R$. Also, $p^{-1}ep$ is an idempotent for every invertible $p \in R$.

Observations from the previous paragraph are meaningless if e is a **central idempotent**, i.e., an idempotent from the center $Z(R)$ of R. In this case $eRf = fRe = 0$ and the Peirce decomposition reduces to two summands $I := eR = eRe$ and $J := fR = fRf$. Clearly I and J are ideals of R, $R = I \oplus J$, and $R \cong I \times J$ via the isomorphism $x \mapsto (ex, fx)$. A central idempotent thus gives rise to a decomposition of a ring.

We have assumed at the beginning that R was unital, which made it possible for us to introduce the idempotent $f = 1 - e$. The presence of 1 could actually be avoided without harming the essence of the above discussion. We did not really deal with f itself, but with products of f with elements from R. Now, if one does not assume that R is unital and then writes $x - ex$ instead of fx, $x - ex - xe + exe$ instead of fxf, etc., then most of what was said above makes sense. In particular, the Peirce decomposition is still available, but the price we have to pay for not assuming the existence of 1 are lengthy and messy formulas. However, the assertions concerning central idempotents need only minor changes. The following can be extracted from the discussion in the previous paragraph: If e is a central idempotent in R, then $I = eR$ and $J = \{x - ex \mid x \in R\}$ are ideals of R such that

$$R = I \oplus J \cong I \times J.$$

Considering I as a ring, we see that e is its unity. Ideals are rarely unital rings. In fact, only those that are generated by central idempotents are.

Lemma 2.54 *Let I be an ideal of a ring R. If I is a unital ring, then its unity e is a central idempotent in R, $I = eR$, and there exists an ideal J of R such that $R = I \oplus J$. Moreover, $R \cong I \times J$.*

Proof Since $e \in I$, we have $eR \subseteq I$, and conversely, $I = eI \subseteq eR$. Thus $I = eR$. Next, since $ex, xe \in I$ for every $x \in R$, we have $ex = (ex)e$ and $xe = e(xe)$. Consequently, $ex = xe$, and so e is a central idempotent. Now we refer to the discussion before the lemma.   □

Again assuming that R is unital, we can state a kind of converse to Lemma 2.54: If I and J are ideals of R such that $R = I \oplus J$, then there exists a central idempotent $e \in R$ such that $I = eR$ and $J = (1 - e)R$ (so I and J are unital rings). The proof is easy. Just write $1 = e + f$ with $e \in I, f \in J$, and check that e has the desired properties.

2.8 Minimal Left Ideals

When considering one-sided ideals we will usually give preference to the left ones.

Definition 2.55 A left ideal L of a ring R is called a **minimal left ideal** if $L \neq 0$ and L does not properly contain a nonzero left ideal of R.

Minimal right ideals and **minimal two-sided ideals** are defined analogously.

Example 2.56 The only minimal left ideal of a division ring D is D itself.

Example 2.57 Let $R = M_n(D)$, D a division ring, and let L be the set of matrices in R that have arbitrary entries in the ith column and zeros in all other columns. It is an easy exercise to show that L is a minimal left ideal of R. We also remark that $L = RE_{ii}$ where E_{ii} is the standard matrix unit, and that $E_{ii}RE_{ii} = \{dE_{ii} \mid d \in D\}$ is a division subring of R isomorphic to D.

Lemma 2.58 *If L is a minimal left ideal of a semiprime ring R, then there exists an idempotent $e \in R$ such that $L = Re$ and eRe is a division ring.*

Proof Since R is semiprime, there exist $x, y \in L$ such that $xy \neq 0$. In particular, $Ly \neq 0$. But Ly is a left ideal of R contained in L, so $Ly = L$ because of the minimality of L. Accordingly, there exists $e \in L$ such that $ey = y$. Hence it follows that $e^2 - e$ belongs to the set $J := \{z \in L | zy = 0\}$. Clearly, J is again a left ideal of R contained in L. Since $x \in L \setminus J$, this time we conclude that $J = 0$. In particular, $e^2 = e$. As $e \in L$, we have $Re \subseteq L$, and since $0 \neq e \in Re$ it follows from the minimality assumption that $L = Re$. Now consider the corner ring eRe. Let $a \in R$ be such that $eae \neq 0$. We must prove that eae is invertible in eRe. We have $0 \neq Reae \subseteq Re = L$, and so $Reae = L$. Therefore $beae = e$ holds for some $b \in R$, and hence also $(ebe)(eae) = e$. Since ebe is a nonzero element in eRe, by the same argument there exists $c \in R$ such that $(ece)(ebe) = e$. But a left inverse coincides with a right inverse, so $eae = ece$ is invertible in eRe, with ebe being its inverse. $\square$

A similar statement of course holds for minimal right ideals. Also, without any change in the proof we see that the lemma holds for algebras as well. Needless to say, minimal left ideals of algebras are defined in the same way as minimal left ideals of rings.

Semiprime rings with minimal left ideals thus contain particularly nice corner rings. Unfortunately, the existence of minimal left ideals in rings is an exception rather than a rule. However, in nonzero finite dimensional algebras they exist for obvious reasons—just take any nonzero left ideal of minimal dimension. The following result thus follows immediately from (the algebra version of) Lemma 2.58.

Corollary 2.59 *If A is a nonzero finite dimensional semiprime algebra, then there exists an idempotent $e \in A$ such that eAe is a division algebra.*

Corollary 2.59 is exactly what we will need in the next section. However, it would be inappropriate to stop the discussion on minimal left ideals at this point without mentioning the converse of Lemma 2.58.

Corollary 2.60 *The following statements are equivalent for an idempotent e in a semiprime ring R:*

(i) *eRe is a division ring.*
(ii) *Re is a minimal left ideal of R.*
(iii) *eR is a minimal right ideal of R.*

Proof Since (i) is, unlike (ii) and (iii), a left-right symmetric condition, it suffices to prove that it is equivalent to (ii). As we know that (ii) implies (i) by (the proof of) Lemma 2.58, we must only show that (i) implies (ii). Thus, assume that eRe is a division ring, and take a left ideal I of R such that $0 \neq I \subseteq Re$. It is enough to show that $e \in I$, as this clearly yields $I = Re$. Pick $0 \neq u \in I$. By the semiprimeness of R we have $uru \neq 0$ for some $r \in R$. Note that $u = ue$ for $u \in I \subseteq Re$. Consequently, $eru = erue$ is a nonzero element of eRe, and hence there exists $v \in R$ such that $(eve)(eru) = e$. Therefore $e \in Ru \subseteq I$. $\square$

We will continue the study of minimal left ideals later, in Sect. 5.4.

2.9 Wedderburn's Structure Theorems

In 1907, Wedderburn laid down the foundations of the structure theory of noncommutative algebras. Equipped with simple lemmas from the previous sections we are now in a position to prove Wedderburn's theorems. The first one characterizes finite dimensional simple algebras.

Theorem 2.61 (Wedderburn) *Let A be a nonzero finite dimensional algebra. The following statements are equivalent:*

(i) *A is prime.*
(ii) *A is simple.*
(iii) *There exist $n \in \mathbb{N}$ and a division algebra D such that $A \cong M_n(D)$.*

Proof We already know that (iii) implies (ii) (Example 1.10), and trivially (ii) implies (i). Thus we only have to prove that (i) implies (iii). Let us first do this under the additional assumption that A is *unital*. The proof is by induction on $d := [A : F]$.

If $d = 1$, then we simply take $n = 1$ and $D = F$. Thus, let $d > 1$. By Corollary 2.59 there exists an idempotent $e \in A$ such that eAe is a division algebra. If $e = 1$, then the desired result holds (with $n = 1$). Assume therefore that e is a nontrivial idempotent. Set $f := 1 - e$. Note that fAf is a nonzero prime algebra with unity f. Since e does not belong to fAf, we have $[fAf : F] < d$. The induction assumption implies that fAf is isomorphic to the algebra of $m \times m$ matrices over some division algebra. Accordingly, fAf contains matrix units e_{ij}, $i, j = 1, \ldots, m$, such that $e_{11}fAfe_{11} = e_{11}Ae_{11}$ is a division algebra. Our goal is to extend the matrix units of fAf to matrix units of A. We begin by setting $n := m + 1$ and $e_{nn} := e$. Then $\sum_{i=1}^{n} e_{ii} = f + e = 1$, and of course $e_{nn}e_{ij} = e_{ij}e_{nn} = 0$ for all $i, j < n$. It remains to find e_{in} and e_{ni}, $i \leq n - 1$ (informally: the last row and the last column of matrices from our algebra are still "unknown", except for their intersection).

Let us first find e_{1n} and e_{n1}. This is the heart of the proof. Using the primeness twice we see that $e_{11}ae_{nn}a'e_{11} \neq 0$ for some $a, a' \in A$. As $e_{11}Ae_{11}$ is a division algebra with unity e_{11}, there is $a'' \in A$ such that

$$(e_{11}ae_{nn}a'e_{11})(e_{11}a''e_{11}) = e_{11}.$$

Setting $e_{1n} := e_{11}ae_{nn}$ and $e_{n1} := e_{nn}a'e_{11}a''e_{11}$ we thus have

$$e_{1n}e_{n1} = e_{11}.$$

Since $e_{n1} \in e_{nn}Ae_{11}$, we have $e_{n1} = e_{nn}e_{n1}$ and $e_{n1} = e_{n1}e_{11} = e_{n1}e_{1n}e_{n1}$. Comparing both relations we get

$$(e_{n1}e_{1n} - e_{nn})e_{n1} = 0.$$

The element $e_{n1}e_{1n} - e_{nn}$ lies in the division algebra $e_{nn}Ae_{nn}$. If it was not zero, then we could multiply the last identity from the left-hand side by its inverse, leading to a contradiction $0 = e_{nn}e_{n1} = e_{n1}$. Therefore

$$e_{n1}e_{1n} = e_{nn}.$$

Finally, we define $e_{nj} := e_{n1}e_{1j}$ and $e_{jn} := e_{j1}e_{1n}$ for $j = 2, \ldots, n - 1$. Note that $e_{ij} = e_{i1}e_{1j}$ and $e_{1j}e_{k1} = \delta_{jk}e_{11}$ holds for all $i, j, k = 1, \ldots, n$. Consequently, for all $i, j, k, l = 1, \ldots, n$ we have

$$e_{ij}e_{kl} = e_{i1}e_{1j}e_{k1}e_{1l} = \delta_{jk}e_{i1}e_{11}e_{1l} = \delta_{jk}e_{i1}e_{1l} = \delta_{jk}e_{il}.$$

Thus e_{ij}, $i, j = 1, \ldots, n$, are indeed matrix units of A. Lemma 2.52 (together with Remark 2.53) yields the desired conclusion $A \cong M_n(D)$, where $D = e_{11}Ae_{11}$.

It remains to show that (i) implies (iii) without the assumption that A is unital. Suppose A is prime and not unital. Then $A^\sharp$ is a prime unital algebra by Lemma 2.36.

Since (i) implies (iii) (and hence (ii)) for unital algebras, it follows that $A^\sharp$ is simple. However, this is a contradiction since A is a proper nonzero ideal of $A^\sharp$. □

We can add to Theorem 2.61 that D is, of course, also finite dimensional, and that $[A : F] = n^2[D : F]$. Let us also mention that the classical version of Theorem 2.61 does not involve prime algebras. Usually it is stated as

$$A \text{ is simple} \iff A \cong M_n(D).$$

However, for our approach to the structure theory the concept of primeness is more suitable than the concept of simplicity. At any rate, it is interesting that finite dimensional prime algebras coincide with simple ones. In general, the class of prime algebras is much larger than the class of simple algebras.

Let us also remark that the restriction to finite dimensions in Theorem 2.61 is absolutely necessary. Just think of the Weyl algebra, which is a simple domain but not a division algebra (see Example 2.28).

Corollary 2.62 *A finite dimensional algebra A is a central simple algebra if and only if there exist $n \in \mathbb{N}$ and a central division algebra D such that $A \cong M_n(D)$.*

Proof Apply Theorem 2.61 and Lemma 1.15. □

Corollary 2.63 *The dimension of a finite dimensional central simple algebra is a perfect square.*

Proof We have $[M_n(D) : F] = n^2[D : F]$. Now use Corollary 1.37. □

The direct product of prime algebras is a semiprime algebra (cf. Example 2.33). We will now show that in the finite dimensional setting every semiprime algebra is built in this way.

Theorem 2.64 (Wedderburn) *Let A be a nonzero finite dimensional algebra. Then A is semiprime if and only if there exist $n_1, \dots, n_r \in \mathbb{N}$ and division algebras $D_1, \dots, D_r$ such that $A \cong M_{n_1}(D_1) \times \cdots \times M_{n_r}(D_r)$.*

Proof We only have to prove the "only if" part. We proceed by induction on $d = [A : F]$. If $d = 1$, then $A = Fa$ with $a^2 \neq 0$. Thus $a^2 = \lambda a$, $\lambda \neq 0$, so that $\lambda^{-1}a$ is the unity of A. Hence $A \cong F$. Let $d > 1$. If A is prime, then the result follows from Theorem 2.61. We may therefore assume that there exists $0 \neq a \in A$ such that $I = \{x \in A \mid aAx = 0\}$ is a nonzero set. Clearly, I is an ideal of A, and therefore a semiprime algebra (cf. Remark 2.25). Since $a \notin I$ we have $[I : F] < d$. The induction assumption implies that $I \cong M_{n_1}(D_1) \times \cdots \times M_{n_p}(D_p)$ for some $n_i \in \mathbb{N}$ and division algebras D_i, $i = 1, \dots, p$. As each factor $M_{n_i}(D_i)$ has a unity, so does I. By Lemma 2.54 there is an ideal J of A such that $A \cong I \times J$. We may use the induction assumption also for J and conclude that $J \cong M_{n_{p+1}}(D_{p+1}) \times \cdots \times M_{n_r}(D_r)$ for some $n_i \in \mathbb{N}$ and division algebras D_i, $i = p + 1, \dots, r$. The result now clearly follows. □

Theorem 2.64 in particular shows that a nonzero finite dimensional semiprime algebra is automatically unital.

A more common name for a finite dimensional semiprime algebra is a (finite dimensional) **semisimple algebra**. Accordingly, a more standard wording of Theorem 2.64 is

$$A \text{ is semisimple} \iff A \cong M_{n_1}(D_1) \times \cdots \times M_{n_r}(D_r).$$

Semisimple rings form another class of rings, which we will meet later. In the finite dimensional context, semisimple algebras coincide with semiprime ones.

If a finite dimensional algebra A has nonzero nilpotent ideals, then we can factor out its radical and thereby obtain a semiprime algebra (Lemma 2.26). This yields the following result which gives a more complete picture of the structure of finite dimensional algebras.

Theorem 2.65 *Let A be a finite dimensional algebra. If N is its radical (i.e., the maximal nilpotent ideal), then*

$$A/N \cong M_{n_1}(D_1) \times \cdots \times M_{n_r}(D_r)$$

for some $n_1, \ldots, n_r \in \mathbb{N}$ and division algebras $D_1, \ldots, D_r$, unless $A = N$ is a nilpotent algebra.

Actually, more can be said. If A is an algebra over a *perfect field* F (fields of characteristic 0, algebraically closed fields, and finite fields are all examples of perfect fields), then there exists a subalgebra A' of A such that $A' \cong A/N \cong M_{n_1}(D_1) \times \cdots \times M_{n_r}(D_r)$ and $A = A' \oplus N$, the vector space direct sum. This result is known as **Wedderburn's principal theorem**. We omit the proof. The interested reader can find it, for instance, in [Row91, pp. 163–164]. For a nice illustration of the theorem examine the case where $A = T_n(F)$, cf. Examples 2.14 and 2.27.

We will give additional information on Wedderburn's structure theorems in Sects. 3.9, 5.3, and 5.4.

2.10 Algebras Over Special Fields

The purpose of this section is to restate Wedderburn's theorems for algebras over some special fields. We begin with algebraically closed fields, for which these theorems take on extremely simple forms. The following result was proved (for $F = \mathbb{C}$) already in 1892 by T. Molien.

Corollary 2.66 (Molien) *Let A be a nonzero finite dimensional semiprime algebra over an algebraically closed field F. Then $A \cong M_{n_1}(F) \times \cdots \times M_{n_r}(F)$ for some $n_1, \ldots, n_r \in \mathbb{N}$. Moreover, if A is prime then $r = 1$.*

Proof Use Theorems 2.61 and 2.64 along with Proposition 1.6.

In the next result we will indicate the usefulness of Corollary 2.66 by combining it with Maschke's theorem. Given a finite group G, we can form the group algebra $F[G]$ for any field F. Understanding $F[G]$ can give us a clue for understanding G. We have a freedom in selecting the field F. In view of Corollary 2.66, choosing F to be algebraically closed seems to be a good idea. Maschke's theorem (Theorem 2.49) suggests that by requiring that char(F) = 0 we will be on the safe side. An obvious choice now is $F = \mathbb{C}$.

Corollary 2.67 *Let G be a finite group. Then there exist $n_1, \ldots, n_r \in \mathbb{N}$ such that $\mathbb{C}[G] \cong M_{n_1}(\mathbb{C}) \times \cdots \times M_{n_r}(\mathbb{C})$.*

Thus, starting with an abstract finite group, we were able to represent it through its group algebra as a "concrete" object, the direct product of full matrix algebras. Corollary 2.67 is a basic result in the representation theory of finite groups. A *representation* of a group G is a group homomorphism from G into the general linear group, i.e., the group of all invertible $n \times n$ matrices.

Example 2.68 Consider the group algebra $\mathbb{C}[S_3]$, where S_3 is the symmetric group of $\{1, 2, 3\}$. Representing it as $M_{n_1}(\mathbb{C}) \times \cdots \times M_{n_r}(\mathbb{C})$, we must have $\sum_{i=1}^{r} n_i^2 = 6$, the dimension of $\mathbb{C}[S_3]$; moreover, at least one n_i must be different from 1 since S_3 is noncommutative. The only possibility is that $\mathbb{C}[S_3] \cong M_2(\mathbb{C}) \times \mathbb{C} \times \mathbb{C}$.

In the next results we confine ourselves to simple algebras.

Corollary 2.69 *A finite dimensional simple $\mathbb{R}$-algebra A is isomorphic either to $M_n(\mathbb{R})$, $M_n(\mathbb{C})$, or $M_n(\mathbb{H})$ for some $n \in \mathbb{N}$.*

Proof Combine Theorem 2.61 with Theorem 1.4.   □

With reference to Corollary 2.62 we get

Corollary 2.70 *A finite dimensional central simple $\mathbb{R}$-algebra A is isomorphic either to $M_n(\mathbb{R})$ or $M_n(\mathbb{H})$ for some $n \in \mathbb{N}$.*

Corollary 2.71 *A finite simple ring R is isomorphic to $M_n(F)$ for some $n \in \mathbb{N}$ and some finite field F.*

Proof As noticed in Remark 2.20, char(R) is a prime number p and we can consider R as a $\mathbb{Z}_p$-algebra. Therefore Theorem 2.61 tells us that $R \cong M_n(D)$ where D is a finite dimensional division algebra over $\mathbb{Z}_p$. By Theorem 1.38, D is a field.   □

Wedderburn's theorems reduce the problem of classifying finite dimensional (semi)prime F-algebras to classifying finite dimensional division F-algebras. The latter is fully understood if F is one of the fields considered in this section. For some fields F, however, finding all finite dimensional division F-algebras may be an extremely difficult problem and their explicit list cannot be given.

2.11 Scalar Extension (A Naive Approach)

In most of the above sections the field F has played just a formal role in our study of F-algebras. In the previous section, however, we saw that F is of essential importance if we wish to describe algebras precisely. We now aim to present a simple construction which enables replacing the given field F by a "better" field for which the structure theory yields definitive results. The standard way for introducing this construction requires the concept of the tensor product. Following the elementary style of this chapter we will use an alternative way which has many disadvantages and one advantage: it is very simple and intuitively clear. It is our hope that a naive approach used here will help the reader to easily understand (and appreciate) the "correct" definition given later, in Sect. 4.7.

Let us begin with an illuminating example.

Example 2.72 Consider the algebra $\mathbb{H}_\mathbb{C}$ of "complex quaternions", i.e., $\mathbb{H}_\mathbb{C}$ is the 4-dimensional $\mathbb{C}$-algebra with basis $\{1, i, j, k\}$ satisfying the same multiplication rules (1.1) as the standard real quaternions $\mathbb{H}$ (cf. Remark 1.7). That is, we keep the multiplication table of $\mathbb{H}$, but change the scalars from $\mathbb{R}$ to $\mathbb{C}$. Every complex algebra can be canonically considered as a real algebra (of dimension $2n$ if it is n-dimensional as a complex algebra). In this sense we may regard $\mathbb{H}$ as an $\mathbb{R}$-subalgebra of $\mathbb{H}_\mathbb{C}$. As a sum of squares of nonzero complex numbers can easily be 0, the proof that $\mathbb{H}$ is a division algebra does not work for $\mathbb{H}_\mathbb{C}$. Moreover, indirectly it shows that $\mathbb{H}_\mathbb{C}$ is not a domain. Let us determine what $\mathbb{H}_\mathbb{C}$ really is. We shall write i for the complex imaginary unit (to distinguish it from $i \in \mathbb{H}_\mathbb{C}$). It is straightforward to check that the elements

$$e_{11} = \frac{1 - ii}{2}, \quad e_{12} = \frac{j - ik}{2}, \quad e_{21} = \frac{-j - ik}{2}, \quad e_{22} = \frac{1 + ii}{2}.$$

form a set of 2×2 matrix units of $\mathbb{H}_\mathbb{C}$, and that $e_{11}\mathbb{H}_\mathbb{C}e_{11} = \mathbb{C}e_{11} \cong \mathbb{C}$. Using Lemma 2.52 we may now conclude that

$$\mathbb{H}_\mathbb{C} \cong M_2(\mathbb{C}).$$

Considering the restriction of this isomorphism to $\mathbb{H}$ we see that $\mathbb{H}$ is isomorphic to the real algebra of all complex matrices of the form $\left[\begin{smallmatrix} z & w \\ -\bar{w} & \bar{z} \end{smallmatrix} \right]$.

One might object that by changing the real scalars into the complex ones we did not make the quaternions nicer. On the contrary, we have spoiled the property of being a division algebra. However, our aim was to indicate a general principle that is hidden behind this example. Let us consider another, admittedly less appealing, in fact quite trivial example.

Example 2.73 The standard matrix units $E_{11}, E_{12}, E_{21}, E_{22}$ form the standard basis of $M_2(\mathbb{R})$. Again substituting the scalars we can consider the $\mathbb{C}$-algebra $M_2(\mathbb{R})_\mathbb{C}$

consisting of elements $\sum_{i,j=1}^{2} \lambda_{ij} E_{ij}$, $\lambda_{ij} \in \mathbb{C}$, whose multiplication is determined by $E_{ij}E_{kl} = \delta_{jk}E_{il}$. It is obvious that

$$M_2(\mathbb{R})_{\mathbb{C}} \cong M_2(\mathbb{C}).$$

Let us summarize what has been inferred in these two examples. From Corollary 2.69 we see that up to isomorphism there are exactly two 4-dimensional simple real algebras: $\mathbb{H}$ and $M_2(\mathbb{R})$ (they both happen to be central). If we take the standard basis of any of them and introduce the algebra that has the same basis and the same multiplication table, only the scalars are changed from $\mathbb{R}$ to $\mathbb{C}$, then this new algebra is isomorphic to $M_2(\mathbb{C})$.

Now we proceed to the abstract situation. Take an algebra A over F. For simplicity assume that it is finite dimensional. Choose a basis $\{e_1, \ldots, e_d\}$ of A with multiplication table

$$e_i e_j = \sum_{k=1}^{d} \alpha_{ijk} e_k. \tag{2.5}$$

The constants $\alpha_{ijk} \in F$ completely determine the multiplication in A. Let K be a field extension of F. We now introduce the K-algebra A_K which has the same basis $\{e_1, \ldots, e_d\}$ and whose multiplication is determined by the same formula (2.5) (this makes sense since $\alpha_{ijk} \in F \subseteq K$). Thus, we only change the scalars but keep the operations. The associativity of the multiplication in A_K is a consequence of the associativity of the products of the e_i's. We call A_K the **scalar extension** of A to K.

Since F is a subfield of K, we can canonically consider A_K as an F-algebra and A as its F-subalgebra. Let us also point out that

$$[A_K : K] = [A : F]. \tag{2.6}$$

Our definition of scalar extension depends on a basis. It would be formally more correct to denote it by something like $A_{K,\{e_1,\ldots,e_d\}}$ instead of just A_K. However, let us put aside a natural question of what happens if we take another basis, and stick with the simplified notation A_K. The "correct" definition that will be given in Sect. 4.7 is independent of basis, and when we reach that point everything should be clear. Let us now only concentrate on a generalization of our observations on 4-dimensional (central) simple real algebras.

Theorem 2.74 *Let A be a finite dimensional central simple F-algebra. If $\overline{F}$ is an algebraic closure of F, then $A_{\overline{F}} \cong M_n(\overline{F})$ where $n = \sqrt{[A : F]}$.*

Proof Let us show that $A_{\overline{F}}$ is simple. Take a nonzero ideal I of $A_{\overline{F}}$, and pick a nonzero $u = \lambda_1 e_1 + \cdots + \lambda_d e_d \in I$. We may assume that $\lambda_1 \neq 0$. Let $\varphi \in \operatorname{End}_F(A)$ be such that $\varphi(e_1) = 1$ and $\varphi(e_i) = 0$, $i \geq 2$. By Lemma 1.25 there exist $a_i, b_i \in A$ such that $\varphi = \sum_{i=1}^{r} L_{a_i} R_{b_i}$. Therefore

$$I \ni \sum_{i=1}^{r} a_i u b_i = \sum_{j=1}^{d} \lambda_j \left(\sum_{i=1}^{r} a_i e_j b_i \right) = \sum_{j=1}^{d} \lambda_j \varphi(e_j) = \lambda_1.$$

Thus I contains an invertible element, so that $I = A_{\overline{F}}$. This means that $A_{\overline{F}}$ is simple. From Corollary 2.66 it follows that $A_{\overline{F}} \cong M_n(\overline{F})$ for some $n \in \mathbb{N}$; (2.6) tells us that $[A : F] = n^2$. ☐

Theorem 2.74 can be used to reduce various problems on finite dimensional central simple algebras to matrix algebras over fields (even better, over algebraically closed fields). A quick illustration: Theorem 2.74 immediately yields a new proof of Corollary 2.63.

Exercises

2.1. Describe the radical of the subalgebra of $M_n(F)$ consisting of all matrices whose last k rows are zero.
2.2. Find an ideal of the algebra $\Pi_{n\geq 2} T_n(F)$ that is nil but not nilpotent.
2.3. Let R be an arbitrary ring. Show that:

(a) The left ideal generated by $a \in R$ is nil if and only if the right ideal generated by a is nil.
(b) The sum of all nil left ideals of R is equal to the sum of all nil right ideals of R, and is therefore an ideal of R.
(c) The sum of a nil left ideal of R and a nil ideal of R is nil.
(d) The sum of all nil ideals of R, Nil(R), is a nil ideal of R, and the ring $R/\text{Nil}(R)$ has no nonzero nil ideals.
(e) If $R/\text{Nil}(R)$ is commutative (in particular, if R itself is commutative), then Nil(R) is equal to the set of all nilpotent elements in R.

Remark: We call Nil(R) the **upper nilradical** of R. There is also the notion of the **lower nilradical** (also called the **prime radical**) of R which is defined as the intersection of all **prime ideals** of R, i.e., ideals with the property that the corresponding factor rings are prime. It can be shown that the lower nilradical is a nil ideal, and hence contained in Nil(R), and that a ring is semiprime if and only if its lower nilradical is zero.

2.4. Prove that the following statements are equivalent:

(i) Köthe's problem has a positive answer.
(ii) The upper nilradical of an arbitrary ring contains all nil one-sided ideals of this ring.
(iii) The sum of two nil left ideals of an arbitrary ring is nil.

2.5. Show that a prime ring without a nonzero nilpotent element is a domain.

2.6. Show that a semiprime ring R is prime if and only if the intersection of any two nonzero ideals of R is nonzero. Give an example showing that this is not always true if R is not semiprime.

2.7. Show that a ring R is prime (resp. semiprime) if and only if $M_n(R)$ is prime (resp. semiprime).

2.8. Let a, b be elements in a ring R. Note that $aRb = 0$ implies $bRa = 0$ in case R is semiprime, and give an example showing that this is not true in every ring R.

2.9. Let R be a prime ring. Show that if $a \in R$ commutes with every element from a nonzero left ideal L of R, then $a \in Z(R)$. Is this still true if R is semiprime?

2.10. Let R be a semiprime ring, and let $a \in R$ be such that $axa = a^2x$ for every $x \in R$. Show that $a \in Z(R)$.

2.11. Find an example of a commutative domain without unity such that its unitization is not a domain.

Remark: Lemma 2.36 thus does not hold for rings.

2.12. Show that a unital F-algebra B is isomorphic to the unitization $A^\sharp$ of some algebra A if and only if there exists a nonzero algebra homomorphism from B into F. Notice that every group algebra $F[G]$ is an example of such an algebra.

2.13. Let A be a nonzero finite dimensional unital algebra over a field F with char$(F) = 0$, and let φ be an automorphism of A. Show that $\varphi(a) \neq a + 1$ for every $a \in A$.

Hint: Express $L_{\varphi(a)}$ through φ, φ^{-1}, and L_a.

2.14. Let A be a unital algebra, and let G be a subgroup of the multiplicative group A^*. Explain why, in general, the group algebra of G is not isomorphic to the subalgebra of A generated by G (which is just the linear span of G, so the elements in both algebras look the same), and find an example (with $G \neq \{1\}$) where these two algebras are isomorphic.

2.15. Show that every group algebra $F[G]$ has a (linear) involution.

2.16. Show that 0 is the only idempotent in a domain without unity.

2.17. Let e_1, e_2 be idempotents in an F-algebra A. Note that if char$(F) \neq 2$, then $e_1 + e_2$ is an idempotent if and only if e_1 and e_2 are orthogonal. Show that this can be extended to any number of idempotents in case A is finite dimensional and char$(F) = 0$; that is, the sum of idempotents $e_1, \ldots, e_n \in A$ is an idempotent if and only if $e_1, \ldots, e_n$ are pairwise orthogonal.

Hint: The trace of an idempotent matrix is equal to its rank.

2.18. Show that the subring generated by all idempotents in a ring R contains the ideal generated by all commutators $[e, x]$ where e is an idempotent and x is an arbitrary element in R. Hence conclude that a simple ring containing at least one nontrivial idempotent is generated (as a ring) by its idempotents.

Hint: Express $[e, x]$ as a difference of two idempotents (an inspection of Sect. 2.7 should give an idea how this can be done), and use the identity $yu = [e, y[e, u]] - [e, y][e, u]$ where $u = [e, x]$.

2.19. Show that a unital ring R contains an element which is right invertible but not left invertible if and only if $M_2(R)$ contains an invertible upper triangular matrix whose inverse is not upper triangular.

Hint: If $ab = 1$, then $e := 1 - ba$ is an idempotent and $ae = eb = 0$.

2.20. Prove that a nonzero idempotent e in a ring R cannot be written as a sum of two nonzero orthogonal idempotents in R if and only if there does not exist an idempotent $f \in R$ different from 0 and e such that $f = ef = fe$. Such an idempotent e is said to be **primitive**. Show that a sufficient condition for e being primitive is that Re is a minimal left ideal of R, and that this condition is also necessary if $R = A$ is a finite dimensional semiprime algebra.

2.21. Let $R_1, \dots, R_n$ be unital rings. Show that every ideal of $R_1 \times \cdots \times R_n$ is of the form $I_1 \times \cdots \times I_n$ where I_i is an ideal of R_i. Hence conclude that the number of ideals of a finite dimensional semiprime algebra is a power of 2.

2.22. Let A be a nonzero finite dimensional algebra. Show that:

   (a) A is nil if and only if A is nilpotent.
   (b) A is a domain if and only if A is a division algebra.
   (c) A has no nonzero nilpotent elements if and only if A is isomorphic to a direct product of division algebras.
   (d) There exists $n \in \mathbb{N}$ such that $x^n = x$ for every $x \in A$ if and only if A is isomorphic to a direct product of finite fields.

   Remark: Later, in Sect. 5.10, we will consider the condition $x^n = x$ in arbitrary rings, even under the assumption that n may depend on x.

2.23. Show that $F[\mathbb{Z}_n] \cong F[\omega]/(\omega^n - 1)$ for every field F. Use this to find the Wedderburn decomposition (in the sense of Theorem 2.64) of $\mathbb{R}[\mathbb{Z}_3]$.

   Remark: This can be also used for describing $\mathbb{C}[\mathbb{Z}_3]$; however, Corollary 2.67 readily implies that $\mathbb{C}[G] \cong \mathbb{C} \times \cdots \times \mathbb{C}$ for every finite abelian group G.

2.24. The subset $Q_8 := \{1, -1, i, -i, j, -j, k, -k\}$ of $\mathbb{H}$ is clearly a group under multiplication. It is called the **quaternion group**. Find the Wedderburn decomposition of $\mathbb{C}[Q_8]$.

   Hint: What is the codimension of the augmentation ideal of a group algebra?

2.25. Let A be a finite dimensional central simple F-algebra such that $[x, y][z, w] + [z, w][x, y] \in F$ for all $x, y, z, w \in A$. Prove that either $A = F$ or $[A : F] = 4$.

Chapter 3
Modules and Vector Spaces

Modules at last! In most books on a similar subject modules appear on the very first pages. And there are good reasons for this. Ring theory and module theory are symbiotic. Many questions about rings can be handled most effectively by using modules. Moreover, sometimes modules seem to be inevitable. In spite of that, we have decided to postpone introducing modules up until now. Our viewpoint has been that an intrinsic approach from the first chapters, where results on rings and algebras were rather quickly derived from basic definitions, should be easier and more appealing for a newcomer to the subject. We have now reached the point where continuing without modules could be inefficient.

This chapter is mostly of a preliminary nature. Several fundamental module-theoretic notions will be considered at an introductory level. The emphasis will be on vector spaces and simple modules. Compared with the previous two chapters, this one lacks striking theorems and surprising proofs. An exception is perhaps a supplement to Wedderburn's structure theory, given close to the end of the chapter.

3.1 Concept of a Module

Being familiar with vector spaces, the definition of a module can be easily memorized. One just replaces the role of a field by an arbitrary ring.

Definition 3.1 Let R be a ring. A **left R-module** (or left module over R) is an additive group M together with a map from $R \times M$ to M, $(r, m) \mapsto rm$, such that for all $r, s \in R$, $m, n \in M$:

(a) $(r + s)m = rm + sm$,
(b) $r(m + n) = rm + rn$,
(c) $r(sm) = (rs)m$.

If, additionally, R is a unital ring and

(d) $1m = m$

for every $m \in M$, then M is said to be a **unital left R-module**.

© Springer International Publishing Switzerland 2014
M. Brešar, *Introduction to Noncommutative Algebra*, Universitext,
DOI 10.1007/978-3-319-08693-4_3

A **(unital) right R-module** is defined similarly, only the order of elements from R and M has to be interchanged. That is, one requires the obvious analogues of (a)–(d) for a map $(m, r) \mapsto mr$, where $m \in M$ and $r \in R$.

Remark 3.2 If R is commutative, then every left R-module M becomes a right R-module by defining $mr := rm$ for all $m \in M, r \in R$.

Distinguishing between left and right R-modules is therefore more substantial when R is noncommutative. This does not mean, of course, that the theory of left R-modules is any different from the theory of right R-modules. Results on left modules have obvious analogues for right modules, and vice versa. Among two equivalent choices, let us decide to work with the left version. Thus, unless specified otherwise, we will henceforth adopt the following convention:

$$R\text{-module} := \text{left } R\text{-module.}$$

When it will be clear which ring R we have in mind, or when R will play just a formal role in our discussion, we will simply use the term "module".

Modules appear all over the place; finding examples is easy, making a selection is harder.

Example 3.3 A vector space over a field F is a unital F-module.

Example 3.4 Every additive group M is a unital $\mathbb{Z}$-module, with nm, where $n \in \mathbb{Z}$ and $m \in M$, having the usual meaning.

Example 3.5 Let M be an additive group. The set $R = \text{End}(M)$ of all endomorphisms of M (i.e., all additive maps from M into M) becomes a ring if we define the sum $f + g$ and the product fg of $f, g \in R$ as follows:

$$(f + g)(m) := f(m) + g(m), \quad (fg)(m) := f(g(m)). \tag{3.1}$$

We can view M as a unital R-module via $fm := f(m), f \in R, m \in M$.

Example 3.6 A left ideal L of a ring R is a left R-module with $r\ell$, where $r \in R$ and $\ell \in L$, simply being the ordinary product in R. In particular, R itself is an R-module.

Similarly, a right ideal of R can be considered as a right R-module, while an ideal (in particular, R itself) is both a left and a right R-module. Moreover, it is an R-bimodule:

Definition 3.7 Let R and S be rings. If M is both a left R-module and a right S-module, and also

$$(rm)s = r(ms) \quad \text{for all } r \in R, m \in M, s \in S,$$

then M is said to be an (R, S)**-bimodule**. An (R, R)-bimodule is also called an R-**bimodule**.

Example 3.8 If R is a subring of a ring R', then R' is an R-bimodule (via the ordinary product in R').

Replacing rings by algebras one has to make some obvious modifications in the above definitions (as well as in examples). For instance, a **(left)** A-**module**, where A is an algebra over a field F, is defined as a vector space V over F together with a map $A \times V \to V$, $(a, v) \mapsto av$, that satisfies (a), (b), (c), and

$$(\lambda a)v = a(\lambda v) = \lambda(av) \quad \text{for all } \lambda \in F, a \in A, v \in V.$$

In other words, the map must be bilinear, not only biadditive.

Example 3.9 The n-dimensional space F^n can be viewed as a unital $M_n(F)$-module if we write vectors in F^n as columns and define the module operation as the standard product of a matrix with a column vector.

Example 3.10 A vector space V over F is a unital $\text{End}_F(V)$-module.

Note that Example 3.10 is a refined version of Example 3.5 which also includes the scalar multiplication, and that Example 3.9 is just a disguised form of its finite dimensional version. The relevance of these examples will become evident from the following discussion.

Let A be an algebra over F and let V be a vector space over F. An algebra homo-morphism from A to $\text{End}_F(V)$ is called a **representation** of A. If $\varphi : A \to \text{End}_F(V)$ is a representation, then V becomes an A-module by defining $av := \varphi(a)(v)$. Con-versely, every A-module V gives rise to a representation $\varphi : A \to \text{End}_F(V)$ given by $\varphi(a)(v) := av$. We may therefore regard representations and modules as equiv-alent concepts. For example, instead of with the regular representation we can deal with the (left) A-module A; thus, in a hidden manner modules were already treated before. We will usually present our results in the language of modules rather than in that of representations. Still, being aware of a possible interpretation in terms of representations is important for understanding the philosophy of the subject. Con-sidering a module over an algebra is essentially the same as representing algebra elements as linear operators of a vector space. This indicates why modules can be useful for the structure theory. Namely, they can be used to identify some abstract algebras with algebras of linear operators. In the previous chapter we have shown that some abstract algebras can be identified with algebras of matrices. But from linear algebra we know that the two seemingly quite different notions, matrices and linear operators, are basically equivalent. Thus, as it turns out, constructing matrix units, the approach taken in the previous chapter, is not the only way to describe algebra elements as matrices—one can rely on modules instead. Perhaps this sounds very vague at this point, especially since the matrices from the previous chapter were not the "ordinary" matrices we know from linear algebra, but matrices with entries from a division algebra. However, as will become apparent in Sect. 3.4, this is not really a problem. In Chap. 5 we will show that the module-theoretic approach to the structure theory yields a profound generalization of Wedderburn's theory.

Similar observations as for modules over algebras can be made for modules over rings: an R-module M can be identified with a ring homomorphism from R into $\text{End}(M)$, the ring of endomorphisms of the additive group of M. Its kernel will later turn out to be important, so we give it a name.

Definition 3.11 The **annihilator** of an R-module M is the set

$$\text{ann}_R(M) := \{r \in R \mid rM = 0\}.$$

If $\text{ann}_R(M) = 0$, then M is said to be a **faithful** module (here, 0 stands for $\{0\}$ and rM for $\{rm \mid m \in M\}$).

A glance at the definition shows that $\text{ann}_R(M)$ is a left ideal, but actually it is—as a kernel of a ring homomorphism—a two-sided ideal.

Example 3.12 Consider the abelian group $\mathbb{Z}_n$ as a $\mathbb{Z}$-module. Then $\text{ann}_{\mathbb{Z}}(\mathbb{Z}_n) = n\mathbb{Z}$.

Example 3.13 The modules from Examples 3.9 and 3.10 are faithful.

3.2 Basic Module-Theoretic Notions

Even the readers that have never met modules before might experience a sense of déjà vu when reading this section. Basic notions of module theory are very similar to those one encounters in the first meetings with vector spaces, groups, and rings. We will therefore introduce them in a rather quick and perhaps slightly dry manner. The reader is advised to translate these notions into the (presumably more familiar) language of vector spaces.

For simplicity of exposition we will consider only modules over rings. It should be obvious how to make the necessary adjustments for modules over algebras. One just has to require the natural scalar multiplication properties at appropriate places.

Throughout this section, R stands for a ring and M for an R-module. Recall that a module means a left module by our convention.

Definition 3.14 A subset L of M is called a **submodule** of M if L is an additive subgroup of M and $r\ell \in L$ for all $r \in R$, $\ell \in L$.

Of course, a submodule is itself a module.

Example 3.15 Every module M has at least two submodules: 0 and M. Submodules different from M are said to be *proper*, and submodules different from 0 are said to be *nonzero* or *nontrivial*.

Example 3.16 If we consider an additive group as a $\mathbb{Z}$-module, then its submodules are exactly its subgroups. Similarly, viewing a vector space over F as an F-module, we see that its submodules are its subspaces.

Example 3.17 The submodules of the R-module R are the left ideals of R. This simple fact will be used frequently.

Example 3.18 Let J be a left ideal of R. The additive subgroup of M generated by the elements of the form um, where $u \in J$ and $m \in M$, is a submodule. We denote it by JM.

The intersection of submodules is obviously a submodule. Given a subset X of M, the intersection of all submodules containing X is therefore the submodule generated by X. Explicitly, it can be described as the additive subgroup generated by the elements $rx + nx$, $r \in R$, $n \in \mathbb{Z}$, $x \in X$ (there is no need to involve nx if M is unital). If a (sub)module N is generated by some finite set, then we say that N is **finitely generated**. Modules generated by only one element are called **cyclic**. A unital cyclic module can be written as

$$Rm := \{rm \mid r \in R\},$$

where m is its generator.

We proceed to another fundamental notion.

Definition 3.19 Let N be another R-module. A map $\varphi : M \to N$ is an **R-module homomorphism** if

$$\varphi(m + m') = \varphi(m) + \varphi(m') \quad \text{and} \quad \varphi(rm) = r\varphi(m)$$

for all $r \in R$ and $m, m' \in M$.

The **kernel** of φ, $\ker \varphi = \{m \in M \mid \varphi(m) = 0\}$, is clearly a submodule of M, and the **image** of φ, $\operatorname{im} \varphi = \{\varphi(m) \mid m \in M\}$, is a submodule of N. An **R-module isomorphism** is, of course, a bijective R-module homomorphism. We say that M and N are **isomorphic modules**, and write $M \cong N$, if there exists an isomorphism $\varphi : M \to N$. It is immediate to check that in this case its inverse $\varphi^{-1} : N \to M$ is also an isomorphism. An **R-module endomorphism** is an R-module homomorphism from M into itself, and an **R-module automorphism** is an endomorphism which is also an isomorphism.

We define the **endomorphism ring**

$$\operatorname{End}_R(M)$$

as the set of all R-module endomorphisms of M endowed with operations defined as in (3.1), i.e., pointwise addition and composition as multiplication. It is immediate to check that $\operatorname{End}_R(M)$ is indeed a (unital) ring. Note that $f \in \operatorname{End}_R(M)$ is invertible if and only if f is a bijective map, i.e., f is an automorphism.

After being familiar with factor groups and factor rings, the following notion should be easy to grasp. Take a submodule L of M. As L is, in particular, a subgroup of the additive group of M, one can form the factor group M/L. Recall that its elements are cosets $m + L = \{m + \ell \mid \ell \in L\}$, $m \in M$, which are added according to

$$(m + L) + (m' + L) = (m + m') + L.$$

We claim that M/L becomes an R-module if we set

$$r(m + L) := rm + L, \quad r \in R, \ m \in M.$$

Indeed, this operation is easily seen to be well-defined, and the module axioms are clearly fulfilled.

Definition 3.20 The R-module M/L is called the **factor module**.

The map $m \mapsto m + L$ is an R-module homomorphism from M onto M/L, called the **canonical homomorphism**. Note that its kernel is equal to L. Another fundamental fact connecting homomorphisms and submodules is the following: If $\varphi : M \to N$ is an R-module homomorphism, then

$$M/\ker \varphi \cong \operatorname{im} \varphi.$$

Indeed, the map $m + \ker \varphi \mapsto \varphi(m)$ is well-defined and is an isomorphism.

Our last theme is about sums and products of (sub)modules.

Definition 3.21 If $\{M_i \mid i \in I\}$ is a family of submodules of M, then the submodule generated by $\bigcup_{i \in I} M_i$ is called the **sum of the submodules** $\{M_i \mid i \in I\}$. We write it as $\sum_{i \in I} M_i$. If every M_j has trivial intersection with the sum of the other submodules, i.e., $M_j \cap \left(\sum_{i \in I \setminus \{j\}} M_i\right) = 0$, then we say that this sum is a **direct sum**, and write it as $\oplus_{i \in I} M_i$.

Note that $\sum_{i \in I} M_i$ consists of all (finite) sums of elements from $\bigcup_{i \in I} M_i$. Hence we see that the sum $\sum_{i \in I} M_i$ is direct if and only if for all distinct $i_1, \dots, i_n \in I$ and all $m_{i_1} \in M_{i_1}, \dots, m_{i_n} \in M_{i_n}$, $m_{i_1} + \cdots + m_{i_n} = 0$ implies $m_{i_1} = \cdots = m_{i_n} = 0$. If the index set I is finite, say $I = \{1, \dots, n\}$, then we can write the sum $\sum_{i \in I} M_i$ as $M_1 + \cdots + M_n$, or $M_1 \oplus \cdots \oplus M_n$ if it is a direct sum.

A formally more adequate term for the above defined direct sum is the **internal direct sum**. One can, on the other hand, start with an arbitrary family of R-modules $\{M_i \mid i \in I\}$, not necessarily submodules of the given module, and then form their **direct product** $\Pi_{i \in I} M_i$: this is the Cartesian product of the M_i's endowed with the componentwise operations

$$(m_i) + (m_i') = (m_i + m_i') \quad \text{and} \quad r(m_i) = (rm_i)$$

for all $r \in R$, $m_i, m_i' \in M_i$, $i \in I$. Its submodule consisting of all elements for which all but finitely many of the components are 0 is called the **external direct sum** of the M_i's. The external direct sum obviously contains isomorphic copies of the modules M_i as its submodules, and, moreover, it is equal to their internal direct sum. Therefore we often neglect the difference between internal and external direct sums, and use the same notation for both. Thus, $M_1 \oplus \cdots \oplus M_n$ may mean an internal or an external direct sum. Since the index set is finite here, this sum coincides with the direct product $\Pi_{i=1}^{n} M_i$ (written also as $M_1 \times \cdots \times M_n$).

Given a submodule L of M, does there exist another submodule K such that $M = L \oplus K$? It is easy to see that the answer is negative in general. Say, the proper nonzero submodules of the $\mathbb{Z}$-module $\mathbb{Z}$ are $n\mathbb{Z}$, $n \geq 2$, and any two of them intersect nontrivially. Submodules for which the answer to our question is positive therefore deserve attention.

Definition 3.22 A submodule L of M is called a **direct summand** of M if there exists a submodule K of M such that $M = L \oplus K$.

Example 3.23 Let L be a left ideal of a unital ring R. Consider L as a submodule of the R-module R. We claim that L is a direct summand if and only if there exists an idempotent $e \in R$ such that $L = Re$. The "if" part is obvious: if $L = Re$ then one easily checks that $K = R(1 - e)$ satisfies $R = L \oplus K$. To prove the converse, suppose there exists a left ideal K such that $R = L \oplus K$. Then there exist $e \in L$ and $f \in K$ such that $1 = e + f$. For every $\ell \in L$ we thus have $\ell - \ell e = \ell f \in L \cap K = 0$, so that $\ell = \ell e$. In particular, $e = e^2$ and $L \subseteq Re$. Since $e \in L$ it follows that $L = Re$.

3.3 Vector Spaces Over Division Rings

The following familiar notion from linear algebra can be also considered in modules.

Definition 3.24 A subset B of an R-module M is said to be **linearly independent** if for all distinct elements $b_1, \ldots, b_n \in B$ and all elements $r_1, \ldots, r_n \in R$, $r_1 b_1 + \cdots + r_n b_n = 0$ implies $r_i = 0$ for every i. If B is not linearly independent, then we say that it is **linearly dependent**.

One should not be misled by the experience with this concept in vector spaces. For instance, if M is not faithful, then even singletons $\{x\}$ are never linearly independent.

Definition 3.25 A linearly independent subset B of an R-module M is called a **basis** of M if M is generated by B.

Assume that R is unital. A unital R-module is said to be a **free R-module** if it has a basis. (A remark for the readers familiar with categories: free R-modules are free objects in the category of unital R-modules.) Thus, a unital R-module M is free if it contains a linearly independent subset B with the following property: for every $m \in M$ there exist $b_1, \ldots, b_n \in B$ and $r_1, \ldots, r_n \in R$ such that $m = r_1 b_1 + \cdots + r_n b_n$. Being free is a rather exceptional property for an R-module, but examples can be easily found. The simplest one is the R-module R. Indeed, $\{1\}$ is its basis. It is easy to see that every free R-module is isomorphic to a direct sum of copies of the R-module R. The structure of free modules is thus quite simple, apparently just as that of vector spaces over fields. Yet a closer look shows that not everything we know about vector spaces holds for free modules. Say, for some rings R it may happen that different bases of a free R-module have different cardinalities. This may arouse the reader's curiosity, but we abandon the discussion on general free modules at this point. A more complete information about them can be found in many other algebra textbooks (for example, [Hun74] gives a gentle introduction). From now on we will consider only free modules that are relevant for our goals. Before introducing them in the next definition, let us point out that up until now it went without saying that "vector space" means a vector space over a field. The following more general conception turns out to be very useful.

Definition 3.26 Let D be a division ring. A unital (left) D-module is called a **(left) vector space** over D.

A right vector space is of course defined as a unital right D-module. Vector spaces over fields are traditionally left vector spaces, but this is just a matter of notational convention (cf. Remark 3.2). One must be more careful about "left" and "right" when considering vector spaces over noncommutative division rings. However, from now on we assume that our vector spaces are left vector spaces.

We believe that the notions such as a **subspace**, **linear combination**, **linear span**, etc., need no explanation. They are defined just as for vector spaces over fields. Fortunately, studying vector spaces over division rings at a basic level is no harder than studying vector spaces over fields. Every vector space is a free module over the corresponding division ring, i.e., the following theorem holds.

Theorem 3.27 *Every vector space has a basis.*

Proof The proof is a typical application of Zorn's lemma. Let $\mathscr{S}$ be the set of all linearly independent subsets of V. Since $\emptyset$ is vacuously a linearly independent set, $\mathscr{S} \neq \emptyset$. Partially order $\mathscr{S}$ by inclusion, i.e., define that $A \leq B$ if $A \subseteq B$. If $\{A_i \mid i \in I\}$ is a chain in $\mathscr{S}$, then it is easy to see that $U := \bigcup_{i \in I} A_i$ is a linearly independent set. Thus $U \in \mathscr{S}$ and obviously it is an upper bound for the chain $\{A_i \mid i \in I\}$. By Zorn's lemma $\mathscr{S}$ contains a maximal element B.

Let us show that B is a basis of V. Since $B \in \mathscr{S}$, it is enough to show that B generates V. If $V = \{0\}$, then $B = \emptyset$ and this is trivially true. Thus, let $V \neq \{0\}$. Pick a nonzero $v \in V$. Our goal is to prove that v is a linear combination of elements from B. We may assume that $v \notin B$. Then $B \cup \{v\}$ contains B as its proper subset, and therefore it is a linearly dependent set. This means that there exists a finite subset of $B \cup \{v\}$ which is linearly dependent. Since subsets of B are linearly independent, this subset must contain v. Thus, there exist $b_1, \ldots, b_n \in B$ and $d_0, d_1, \ldots, d_n \in D$, not all zero, such that
$$d_0 v + d_1 b_1 + \cdots + d_n b_n = 0.$$

The linear independence of $b_1, \ldots, b_n$ implies that $d_0 \neq 0$. Therefore
$$v = (-d_0^{-1} d_1) b_1 + \cdots + (-d_0^{-1} d_n) b_n$$

is indeed a linear combination of elements from B. $\square$

If we introduce $\mathscr{S}$ as the set of all linearly independent subsets of V that contain a given linearly independent set, the same proof gives the following.

Theorem 3.28 *Every linearly independent subset of a vector space is contained in a basis.*

A vector space V is said to be **finite dimensional** if it has a finite basis. Suppose V contains a basis with n elements. Then:

(a) Every basis of V has n elements.
(b) If U is a subspace of V, then its bases have $\leq n$ elements.
(c) If a subspace U of V has a basis with n elements, then $U = V$.

For vector spaces over a field these are the most basic facts that one learns in a standard linear algebra course. The reader must have seen the proofs. Glancing through them one notices that they still work if a field is replaced by a noncommutative division ring. Therefore we feel that we can omit giving these proofs without harming the reader's understanding of the subject (the unconvinced reader can take a look at, for example, [Hun74, pp. 185–187]). In view of (a), we can define the **dimension** of V, $\dim_D V$, to be n if V contains a basis with n elements.

3.4 Endomorphisms and Matrices

Let V be a vector space over a division ring D. The endomorphism ring $\mathrm{End}_D(V)$ will often play an important role in our discussions. Its elements, the endomorphisms, are also called **linear maps** or **linear operators**. Each linear map is uniquely determined by the action on a basis. More precisely, the following holds.

Remark 3.29 Let $B = \{b_i | i \in I\}$ be a basis of V. For each family of vectors $\{w_i \mid i \in I\}$ in V there exists a unique $f \in \mathrm{End}_D(V)$ such that $f(b_i) = w_i$ for every $i \in I$. Indeed, the uniqueness is clear since B is a basis, and the existence is also clear as we may extend f from B to V by linearity, i.e., we define

$$f(d_{i_1}b_{i_1} + \cdots + d_{i_n}b_{i_n}) = d_{i_1}w_{i_1} + \cdots + d_{i_n}w_{i_n}$$

for every finite subset $\{i_1, \ldots, i_n\}$ of I and arbitrary $d_{i_1}, \ldots, d_{i_n} \in D$.

Assume now that V is n-dimensional. If $D = F$ is a field, then we know that the ring $\mathrm{End}_F(V)$ is isomorphic to the matrix ring $M_n(F)$. Representing every endomorphism by a matrix with respect to a fixed basis gives rise to a standard isomorphism. Let us see if this still works if D is a noncommutative division ring.

Choose and fix a basis $\{b_1, b_2, \ldots, b_n\}$ of V. Given $f \in \mathrm{End}_D(V)$, we have

$$f(b_j) = \sum_{i=1}^{n} f_{ij}b_i, \quad j = 1, \ldots, n,$$

for some $f_{ij} \in D$. We attach the $n \times n$ matrix $\varphi(f)$ to f so that f_{ij} is its (i, j) entry, i.e.,

$$\varphi(f) := \begin{bmatrix} f_{11} & f_{12} & \cdots & f_{1n} \\ f_{21} & f_{22} & \cdots & f_{2n} \\ \vdots & \vdots & \ddots & \vdots \\ f_{n1} & f_{n2} & \cdots & f_{nn} \end{bmatrix}.$$

Let g be another element from $\mathrm{End}_D(V)$, and let $g_{ij} \in D$ be such that

$$g(b_j) = \sum_{i=1}^{n} g_{ij} b_i, \quad j = 1, \ldots, n.$$

Then

$$(f + g)(b_j) = \sum_{i=1}^{n} (f_{ij} + g_{ij}) b_i,$$

and hence $\varphi(f + g) = \varphi(f) + \varphi(g)$. If $\varphi(f) = 0$, then f vanishes on a basis, and so $f = 0$. This proves the injectivity of φ. The surjectivity is also obvious in view of Remark 3.29. Indeed, given arbitrary $f_{ij} \in D$, we can define $f \in \mathrm{End}_D(V)$ according to $f(b_j) = \sum_{i=1}^{n} f_{ij} b_i$, so that $\varphi(f) = (f_{ij})$. Thus, φ is an isomorphism of additive groups. This is hardly surprising. Let us now examine the action of φ on products. To determine $\varphi(fg)$, we compute

$$(fg)(b_j) = f(g(b_j)) = \sum_{k=1}^{n} g_{kj} f(b_k) = \sum_{k=1}^{n} g_{kj} \left(\sum_{i=1}^{n} f_{ik} b_i \right) = \sum_{i=1}^{n} \left(\sum_{k=1}^{n} g_{kj} f_{ik} \right) b_i.$$

Thus, the (i, j) entry of $\varphi(fg)$ is

$$\sum_{k=1}^{n} g_{kj} f_{ik}. \tag{3.2}$$

We can now compare this with the (i, j) entry of $\varphi(f)\varphi(g)$ which is equal to

$$\sum_{k=1}^{n} f_{ik} g_{kj}. \tag{3.3}$$

As D may be noncommutative, there is, unfortunately, no reason to believe that (3.2) and (3.3) coincide. But let us look on the bright side: the summations (3.2) and (3.3) are very similar, the terms differ only in the order of factors. This can be remedied. We need the following definition.

Definition 3.30 Let R be a ring. The **opposite ring** of R, denoted by R°, is the ring consisting of the same elements as R, having the same addition as R, and multiplication $\cdot$ given by

$$x \cdot y := yx,$$

where yx is the product in R.

Note that R° is indeed a ring; in particular, the product $\cdot$ is associative. This ring often naturally appears in the study of modules. For example, an obvious adaptation of Remark 3.2 shows that there is a canonical correspondence between left R-modules and right R°-modules.

The rings R and R° are not always isomorphic, but they are **antiisomorphic**. We say that a bijective additive map θ from a ring R onto a ring S is an **antiisomorphism** if $\theta(xy) = \theta(y)\theta(x)$ for all $x, y \in R$. Obviously, the map $x \mapsto x$ is an antiisomorphism from R onto R°. Antiisomorphic rings are simultaneously division rings. Thus D is a division ring if and only if D° is.

Let us now consider $\varphi(f)$ as an element of $M_n(D^{\circ})$ (rather than $M_n(D)$). This point of view does not affect the above discussion up until (3.3). In $M_n(D^{\circ})$, the (i, j) entry of $\varphi(f)\varphi(g)$ is equal to

$$\sum_{k=1}^{n} f_{ik} \cdot g_{kj} = \sum_{j=1}^{n} g_{kj} f_{ik}$$

which does coincide with (3.2), the (i, j) entry of $\varphi(fg)$. Thus φ is an isomorphism from $\mathrm{End}_D(V)$ onto $M_n(D^{\circ})$. We have proved

Theorem 3.31 *Let V be an n-dimensional vector space over a division ring D. Then the rings $\mathrm{End}_D(V)$ and $M_n(D^{\circ})$ are isomorphic.*

Remark 3.32 It is clear from the proof that Theorem 3.31 also holds for free modules. That is, $\mathrm{End}_R(M) \cong M_n(R^{\circ})$ if M is a free R-module having a basis with n elements.

Remark 3.33 From Theorem 3.31 and Example 1.10 we infer that the ring $\mathrm{End}_D(V)$ is simple if V is finite dimensional. (The converse is actually also true, the interested reader can look at Example 5.10).

Remark 3.34 Theorem 3.31 yields an alternative aspect of Wedderburn's structure theory. We now see that we can add another equivalent condition to (i)–(iii) of Theorem 2.61: (iv) $A \cong \mathrm{End}_\Delta(V)$ where Δ is a division algebra (in fact, $\Delta \cong D^{\circ}$) and V is an n-dimensional vector space over Δ. Let us remark that V is automatically also a vector space over the base field F, since $F(= F \cdot 1)$ is contained in Δ. The elements in $\mathrm{End}_\Delta(V)$ are, in particular, F-linear maps.

3.5 Simple Modules

In Sect. 3.3 we have seen that the structure of an R-module is quite simple if $R = D$ is a division ring. We shall now consider another class of easily approachable modules. However, they are of a quite different nature.

Definition 3.35 Let R be a ring. An R-module M is said to be **simple** if $RM \neq 0$ and its only submodules are 0 and M.

If M is unital, then the condition $RM \neq 0$ reduces to $M \neq 0$.

Simple modules are of enormous importance in ring theory. From now on they will frequently play an important role in our discussions. In Chap. 5 we will see that the structure theory of rings rests upon them.

Another term for a simple module is an **irreducible module**. An **irreducible representation** of an algebra A is defined as a representation $\varphi : A \to \mathrm{End}_F(V)$ for which there does not exist a proper nonzero subspace W of V that is invariant under every $\varphi(a)$, $a \in A$. It is easy to check that this is equivalent to the condition that the associated A-module is irreducible.

The following lemma shows that simple modules are cyclic modules generated by *every* nonzero element.

Lemma 3.36 *An R-module $M \neq 0$ is simple if and only if $M = Rm$ for every $0 \neq m \in M$.*

Proof Suppose M is simple. Let $m \in M$. As Rm is a submodule of M, we have either $Rm = 0$ or $Rm = M$. Since $N := \{m \in M \mid Rm = 0\}$ is also a submodule of M and since $RM \neq 0$, we have $N = 0$. Therefore $Rm = M$ whenever $m \neq 0$.

Conversely, if $M = Rm$ for every $0 \neq m \in M$ and L is a nonzero submodule of M, then $M = R\ell \subseteq L$ for every $0 \neq \ell \in L$. Thus $L = M$. $\square$

Viewing modules as generalizations of vector spaces, one might first wonder what does the simplicity mean for the latter. The answer is easy:

Example 3.37 A vector space V over a division ring D is simple (as a D-module) if and only if $\dim_D V = 1$.

This example, however, is not really illuminating. The next ones better reflect the meaning of simple modules in the theory that we shall later develop.

Example 3.38 Let V be a nonzero vector space over a division ring D. As in Example 3.10 we may regard V as a module over $R = \mathrm{End}_D(V)$ (via $fv := f(v)$). We claim that V is a simple R-module. In view of Lemma 3.36 it suffices to show that for each pair $v, w \in V$ with $v \neq 0$ there exists $f \in R$ such that $f(v) = w$. Now, by Theorem 3.28 we may choose a basis of V containing v, and then use Remark 3.29 to conclude that such an f indeed exists.

Example 3.39 Let L be a left ideal of a ring R such that $RL \neq 0$. Then L is simple as an R-module if and only if L is a minimal left ideal.

The following lemma indicates the significance of this example.

Lemma 3.40 *Let R be a simple ring having a minimal left ideal L. Then every simple R-module is isomorphic to L.*

Proof Let M be a simple R-module. By simplicity of R we have $LR = R$. Hence $LM \supseteq L(RM) = (LR)M = RM \neq 0$. Choose $m \in M$ such that $Lm \neq 0$. Since Lm is a submodule of M it follows that $Lm = M$. The map $\varphi : L \to M$ given by $\varphi(\ell) = \ell m$ is therefore a surjective R-module homomorphism. Its kernel is a left ideal of R properly contained in L, so it can only be 0. Therefore φ is an isomorphism. $\square$

Example 3.41 Let $R = M_n(D)$, D a division ring. Then R is a simple ring (Example 1.10). A typical example of a minimal left ideal is RE_{11}, the set of all matrices that have zeros in all columns except in the first one (Example 2.57). Therefore every simple R-module is isomorphic to RE_{11}. Neglecting the zero columns, we may identify this module with D^n, the abelian group of all n-tuple columns on which R acts via the usual matrix multiplication. (A side remark, related to Theorem 3.31: We can consider D^n as a left vector space over D, but then the matrices from R do not act on D^n as D-linear operators; if, however, we consider D^n as a right vector space over D, then they do.)

3.6 Maximal Left Ideals

Minimal left ideals are prototypes of simple modules. However, they only rarely exist in rings. Say, even such a simple and basic example of a ring as $\mathbb{Z}$ does not have them (why?). Let us now consider left ideals having the opposite property. As it turns out, they also have a tight connection with simple modules.

Definition 3.42 A left ideal U of a ring R is called a **maximal left ideal** if $U \neq R$ and U is not properly contained in a proper left ideal.

Maximal right ideals and maximal ideals are defined analogously.

Example 3.43 Let $R = M_n(D)$, D a division ring, and let $U = RE_{11} \oplus \cdots \oplus RE_{n-1,n-1}$. That is, U is the set of all matrices that have zeros in the last column. It is easy to check that U is a maximal left ideal of R. We remark that $R = L \oplus U$ where L is the minimal left ideal RE_{nn}.

Example 3.44 If R is a commutative unital ring, then an ideal U of R is maximal if and only if the factor ring R/U is a field. This is well-known, and the proof is easy (cf. Lemma 5.38 below, which considers a more general situation). Hence:

(a) $p\mathbb{Z}$ is a maximal ideal of $\mathbb{Z}$ for every prime number p.
(b) $U_c = \{f \in C[a, b] \,|\, f(c) = 0\}$, where c is a fixed number from $[a, b]$, is a maximal ideal of the ring of continuous functions $C[a, b]$. Indeed, $f \mapsto f(c)$ is a homomorphism from $C[a, b]$ onto $\mathbb{R}$ with kernel U_c, so that $C[a, b]/U_c \cong \mathbb{R}$ is a field.

It turns out that the U_c's are actually the only maximal ideals of $C[a, b]$. Of course, the $p\mathbb{Z}$'s are the only maximal ideals of $\mathbb{Z}$.

Most rings have plenty of maximal left ideals.

Lemma 3.45 *If R is a unital ring, then every proper left ideal L of R is contained in a maximal left ideal.*

Proof Let $\mathscr{S}$ be the set of all proper left ideals of R that contain L. Since $L \in \mathscr{S}$, $\mathscr{S}$ is nonempty. Partially order $\mathscr{S}$ by inclusion. If $\{K_i \mid i \in I\}$ is a chain in $\mathscr{S}$, then $K = \bigcup_{i \in I} K_i$ is its upper bound. Indeed, K is easily seen to be a left ideal, and $K \neq R$ since $1 \notin K$. By Zorn's lemma $\mathscr{S}$ contains a maximal element, which is obviously a maximal left ideal of R containing L. $\square$

One can imagine that a maximal left ideal U of a ring R is "too big" to be simple as an R-module. However, we will be interested in the factor module R/U which should be expected to be "small". Here we regard R as an R-module and U as its submodule, so the operations in R/U are given by

$$(r + U) + (s + U) = (r + s) + U \quad \text{and} \quad r(s + U) = rs + U.$$

Lemma 3.46 *Let R be a ring. Every simple R-module is isomorphic to R/U for some maximal left ideal U of R. Conversely, if U is a maximal left ideal of R such that $R^2 \not\subseteq U$ (this is automatically fulfilled if R is unital), then R/U is a simple R-module.*

Proof Let M be a simple R-module. Choose $0 \neq m \in M$. Then $Rm = M$ and hence $\varphi : R \to M$, $\varphi(r) = rm$, is a surjective R-module homomorphism. Its kernel U is a left ideal of R, and $R/U \cong M$. We have to show that U is maximal. Let L be a left ideal of R such that $U \subseteq L \subseteq R$. Then L/U is a submodule of R/U. Since R/U, being isomorphic to M, is a simple module, it follows that $L/U = 0$ or $L/U = R/U$. Accordingly, $L = U$ or $L = R$, showing that U is maximal.

To establish the converse, take a maximal left ideal U of R, and consider a submodule N of R/U. Clearly, $L := \{\ell \in R \mid \ell + U \in N\}$ is a left ideal containing U. Thus $L = U$ or $L = R$, yielding $N = 0$ or $N = R/U$. Therefore R/U is simple provided that $R \cdot R/U \neq 0$, i.e., $R^2 \not\subseteq U$. $\square$

Example 3.47 Let R, U, and L be as in Example 3.43. Note that $R/U \cong L$.

Example 3.48 By Example 3.44 it is now clear what are the simple $\mathbb{Z}$-modules and $C[a, b]$-modules.

Lemmas 3.45 and 3.46 yield

Corollary 3.49 *Every nonzero unital ring has a simple module.*

3.7 Schur's Lemma

In the previous chapter we saw that minimal left ideals give rise to division rings (Lemma 2.58), and this was our basis for developing the structure theory of finite dimensional algebras. Simple modules, which can be regarded as generalizations of minimal left ideals (Example 3.39), have a similar role in the structure theory of more general rings and algebras. As a first step towards this more general theory we will now show that simple modules give rise to division rings. The next lemma may be just a straightforward observation, but an extremely useful one. It is named after I. Schur who used it in the representation theory of groups.

Lemma 3.50 (Schur's Lemma) *If M is a simple R-module, then the endomorphism ring $\mathrm{End}_R(M)$ is a division ring.*

Proof Let $f \in \mathrm{End}_R(M)$, $f \neq 0$. Then $\ker f$ is a submodule of M different from M, and $\mathrm{im} f$ is a submodule of M different from 0. Since M is simple, the only possibility is that $\ker f = 0$ and $\mathrm{im} f = M$. Thus f is an automorphism, and therefore an invertible element in $\mathrm{End}_R(M)$. □

A slightly more general version of Schur's Lemma states that a nonzero homomorphism from an arbitrary module into a simple module is surjective, and a nonzero homomorphism from a simple module into an arbitrary module is injective.

A minimal left ideal L of a semiprime ring R thus gives rise to two division rings: the first one is described in Lemma 2.58 and the other one arises from Schur's Lemma if we regard L as a simple R-module. What is the connection between them? The answer will be given in the corollary to the next lemma.

Lemma 3.51 *Let e be an idempotent in a ring R. Then $\mathrm{End}_R(Re) \cong (eRe)^{\circ}$.*

Proof Define $\varphi : \mathrm{End}_R(Re) \to eRe$ by

$$\varphi(f) := f(e).$$

Since f is an endomorphism, we have $f(e) = f(ee) = ef(e)$, and since f maps into Re we have $f(e) = f(e)e$. Accordingly, $f(e) = ef(e)e$ indeed belongs to eRe. The additivity of φ is clear. So is the injectivity: $f(e) = 0$ implies $f(re) = rf(e) = 0$ for every $r \in R$. Pick $a \in eRe$ and define $f : Re \to Re$ by $f(u) := ua$. Then $f \in \mathrm{End}_R(Re)$ and $\varphi(f) = f(e) = a$, which proves that φ is surjective. Next,

$$\varphi(fg) = f(g(e)) = f(g(e)e) = g(e)f(e) = \varphi(g)\varphi(f).$$

Therefore φ is an antiisomorphism. Composing φ with the canonical antiisomorphism $x \mapsto x$ from eRe onto $(eRe)^{\circ}$ we thus obtain an isomorphism from $\mathrm{End}_R(Re)$ onto $(eRe)^{\circ}$. □

Corollary 3.52 *Let L be a minimal left ideal of a semiprime ring R. Then there exists an idempotent $e \in R$ such that $L = Re$, $D = eRe$ is a division ring, and $\mathrm{End}_R(L) \cong D^{\circ}$.*

Proof Use Lemmas 2.58 and 3.51. □

3.8 Semisimple Modules

Theorem 3.27 saying that every vector space has a basis can be equivalently stated as that every vector space is the direct sum of a family of 1-dimensional subspaces. In view of Example 3.37 this can be further rephrased as follows: If D is a division ring, then every unital D-module is the direct sum of a (possibly infinite) family of simple submodules. We will now consider modules with this property over arbitrary rings.

Definition 3.53 Let R be a ring. An R-module M is said to be **semisimple** if M is the direct sum of a family of simple submodules.

The sum of an empty family of submodules should be understood as 0. The zero module is therefore semisimple. As indicated above, all unital modules over division rings are semisimple. Trivially, simple modules are semisimple. A more interesting example, related to Examples 2.57, 3.39 and 3.43, is

Example 3.54 Let $R = M_n(D)$, D a division ring. Pick $1 \leq k \leq n$, and set $M = RE_{11} \oplus \cdots \oplus RE_{kk}$. Thus, M is a left ideal of R consisting of all matrices that have zeros in the last $n - k$ columns. Each RE_{ii} is a minimal left ideal of R, and hence a simple module. Therefore M is a semisimple module. In particular, $R = RE_{11} \oplus \cdots \oplus RE_{nn}$ is semisimple as an R-module.

Recall a familiar fact from linear algebra: Every generating set of a vector space V contains a basis of V. In other words, if V is equal to the sum of 1-dimensional subspaces, then V is equal to the *direct* sum of some of them. The next lemma extends this fact to modules over arbitrary rings.

Lemma 3.55 *If a module M is the sum of a family of simple submodules $\{M_i \mid i \in I\}$, then there exists $K \subseteq I$ such that M is the direct sum of the family $\{M_k \mid k \in K\}$.*

Proof Let $\mathscr{S}$ be the set of all $J \subseteq I$ such that the sum of the family $\{M_j \mid j \in J\}$ is direct. Clearly $\mathscr{S}$ is a nonempty set since $\{i\} \in \mathscr{S}$ for every $i \in I$, and can be partially ordered by inclusion. The existence of a maximal element in $\mathscr{S}$ is a typical application of Zorn's lemma. Indeed, just note that the upper bound of every chain is its union. Denote a maximal element by K, and set $M_0 := \oplus_{k \in K} M_k$. Suppose there exists $i \in I$ such that $M_i \nsubseteq M_0$. Then $M_i \cap M_0$ is a proper submodule of M_i. As M_i is simple, we must have $M_i \cap M_0 = 0$. Accordingly, the sum of the family $\{M_j \mid j \in K \cup \{i\}\}$ is direct, which contradicts the maximality of K. Hence $M_i \subseteq M_0$ for each $i \in I$, so that $M_0 = M$. □

This lemma implies that one can equivalently define semisimple modules by omitting the word "direct" in Definition 3.53:

Corollary 3.56 *If a module is the sum of a family of simple submodules, then it is semisimple.*

Every subspace of a vector space is a direct summand. The reader probably knows this fact at least for finite dimensional vector spaces over fields. For general vector spaces it can be derived from Theorem 3.28. Let us show that it actually holds for semisimple modules.

Lemma 3.57 *Every submodule of a semisimple module is a direct summand.*

Proof Let M be a semisimple module. Thus $M = \oplus_{i \in I} M_i$ where M_i are simple submodules. Let L be a submodule of M. By simplicity of M_i we either have $L \cap M_i = M_i$ or $L \cap M_i = 0$, i.e., L either contains M_i or has trivial intersection with it. If the former holds for every $i \in I$, then $L = M$ and the desired conclusion trivially holds. We may therefore assume that $L \cap M_i = 0$ for some $i \in I$.

As one should expect, we have to invoke Zorn's lemma at one point. Consider the set $\mathscr{S}$ of all $J \subseteq I$ such that $L \cap \left(\oplus_{j \in J} M_j \right) = 0$; $\mathscr{S}$ is nonempty by the previous paragraph. A standard argument shows that $\mathscr{S}$ has a maximal element, let us call it K, with respect to inclusion. We claim that $L \oplus \left(\oplus_{k \in K} M_k \right) = M$. If this was not true, there would exist $i \in I$ such that $M_i \not\subseteq L \oplus \left(\oplus_{k \in K} M_k \right)$, and hence, by simplicity of M_i, $M_i \cap \left(L \oplus \left(\oplus_{k \in K} M_k \right) \right) = 0$. But then $L \cap \left(\oplus_{j \in K \cup \{i\}} M_j \right) = 0$, contradicting the maximality of K. $\square$

It turns out that for unital modules the converse is also true: If every submodule of a unital module M is a direct summand, then M is semisimple. We will not need this result, so we omit the proof. One can find it, for example, in [Row08, p. 36].

Do not be misled with subspaces of vector spaces—submodules are only seldom direct summands. For instance, proper nonzero submodules of the $\mathbb{Z}$-module $\mathbb{Z}$ are $n\mathbb{Z}$, $n \geq 2$, and none of them is a direct summand. Thus $\mathbb{Z}$ is not a semisimple $\mathbb{Z}$-module. We will soon learn that the semisimplicity of the R-module R is in fact an extremely restrictive condition.

3.9 Wedderburn's Structure Theory Revisited

So far we have introduced several concepts related to modules and considered their formal properties. It is about time to experience module theory in action. We are going to do this now by giving a useful supplement to Wedderburn's structure theory.

We will thus consider finite dimensional algebras and modules over them. In the proofs we will refer to previous sections which, however, deal only with modules over rings. This should not confuse the reader. Everything from these sections makes sense for modules over algebras. Of course, all definitions must be appropriately interpreted in the context of algebras. For example, a simple module over an algebra should have no proper nonzero *algebra* submodules, algebra module homomorphisms must be *linear* maps, etc. These changes do not really affect the proofs.

We can add the *uniqueness* statement to Wedderburn's Theorem 2.61.

Theorem 3.58 *If A is a finite dimensional prime (or equivalently, simple) algebra, then there exist a unique $n \in \mathbb{N}$ and an (up to isomorphism) unique division algebra D such that $A \cong M_n(D)$.*

Proof We only have to prove the uniqueness part. Assume therefore that $A \cong M_n(D)$ and $A \cong M_m(C)$ for some $n, m \in \mathbb{N}$ and division algebras D and C. It suffices to prove that $D \cong C$ since this in particular implies that D and C are of the same dimension, from which $n = m$ follows.

Let $\varphi : M_n(D) \to A$ be an isomorphism. Set $e := \varphi(E_{11})$. Since $M_n(D)E_{11}$ is a minimal left ideal of $M_n(D)$, $Ae = \varphi(M_n(D)E_{11})$ is a minimal left ideal of A. By Lemma 3.51, $\mathrm{End}_A(Ae) \cong (eAe)^\circ$. On the other hand, $eAe = \varphi(E_{11}M_n(D)E_{11}) \cong E_{11}M_n(D)E_{11} \cong D$. Accordingly,

$$\mathrm{End}_A(Ae) \cong D^\circ.$$

Analogously, there exists an idempotent $f \in A$ such that Af is a minimal left ideal of A and

$$\mathrm{End}_A(Af) \cong C^\circ.$$

As both Ae and Af are simple A-modules, Lemma 3.40 tells us that they are isomorphic. This implies that the algebras $\mathrm{End}_A(Ae)$ and $\mathrm{End}_A(Af)$ are isomorphic. Indeed, if $\theta : Ae \to Af$ is a module isomorphism, then $g \mapsto \theta g \theta^{-1}$ defines an algebra isomorphism from $\mathrm{End}_A(Ae)$ onto $\mathrm{End}_A(Af)$. Consequently, $D^\circ \cong C^\circ$, and hence $D \cong C$. $\qquad\square$

Wedderburn's Theorem 2.64 can be for unital algebras extended as follows.

Theorem 3.59 *Let A be a nonzero finite dimensional unital algebra. The following statements are equivalent:*

(i) *A is semiprime.*
(ii) *$A \cong M_{n_1}(D_1) \times \cdots \times M_{n_r}(D_r)$ for some $n_i \in \mathbb{N}$ and division algebras D_i.*
(iii) *Every unital A-module is semisimple.*
(iv) *A is a semisimple A-module.*
(v) *Every left ideal of A is of the form Ae where $e \in A$ is an idempotent.*

Proof (i) $\implies$ (ii). Theorem 2.64.

(ii) $\implies$ (iii). We claim that (ii) implies that there exist pairwise orthogonal idempotents $e_1, \ldots, e_s$, where $s = n_1 + \cdots + n_r$, such that $e_1 + \cdots + e_s = 1$, $A = Ae_1 \oplus \cdots \oplus Ae_s$, and each $L_i := Ae_i$ is a minimal left ideal of A. Observe, first of all, that it suffices to show this for $A = M_n(D)$. In this case we have $A = AE_{11} \oplus \cdots \oplus AE_{nn}$ and each AE_{ii} is a minimal left ideal of A (cf. Example 3.54), which proves our claim.

Let M be a unital A-module. For every $m \in M$ we have

$$m = 1m = e_1m + \cdots + e_sm \in L_1m + \cdots + L_sm.$$

Thus, M is the sum of the family of its submodules $\{L_im \mid m \in M, 1 \leq i \leq s\}$. Pick m and i such that $L_im \neq 0$. Let us show that L_im is a simple submodule of M. We will actually show that $L_i m$ is isomorphic to L_i which is, as a minimal left ideal of A, a simple A-module. Clearly, $\ell \mapsto \ell m$ defines an A-module homomorphism from L_i onto $L_i m$. Its kernel is a proper submodule of a simple module L_i, so it must be 0. Therefore $L_i \cong L_i m$ indeed holds. We now see that M is the sum of the family of simple submodules $\{L_i m \mid m \in M, 1 \leq i \leq s, L_i m \neq 0\}$, and thus M is semisimple by Corollary 3.56.

(iii) $\implies$ (iv). Trivial.

(iv) $\implies$ (v). Let L be a left ideal of A. Consider L as a submodule of a semisimple module A. From Lemma 3.57 it follows that L is a direct summand. As shown in Example 3.23, this is equivalent to the existence of an idempotent $e \in A$ such that $L = Ae$.

(v) $\implies$ (i). A left ideal containing a nonzero idempotent cannot be nilpotent. $\square$

Unlike (iii)–(v), (i) and (ii) are left-right symmetric conditions. Therefore we can, in particular, add the following equivalent statement to the above list: (v') Every right ideal of A is of the form eA where e is an idempotent.

Corollary 3.60 *Let A be a finite dimensional simple algebra. If finite dimensional unital A-modules have the same dimension, then they are isomorphic.*

Proof Let M be a unital A-module. By Theorem 3.59 (iii), M is semisimple. If M is also finite dimensional, then it is obviously a finite direct sum of simple modules: $M = M_1 \oplus \cdots \oplus M_r$, M_i simple. By Lemma 3.40, each M_i is isomorphic to a minimal left ideal L of A. Therefore M is isomorphic to the direct sum of r copies of L. The number r thus determines M up to isomorphism. On the other hand, r is determined by the dimension of M. Thus, if N is a unital A-module of the same dimension as M, then $M \cong N$. $\square$

A brief inspection of the proof of Theorem 3.59 shows that we have used the assumption that A is finite dimensional only in the proof of (i) $\implies$ (ii). All other implications hold for every unital algebra, and even for every unital ring (provided that in (ii) we assume that D_i are division rings, not algebras). Certainly (i) $\implies$ (ii) does not hold for general unital rings, there are a myriad of counterexamples ($\mathbb{Z}$ to start with). However, it turns out that (v) $\implies$ (ii) does hold. Thus, (ii)–(v) *are equivalent if $A = R$ is an arbitrary unital ring*. A ring satisfying these equivalent conditions is called a **semisimple ring**. Clearly, a ring is isomorphic to $M_n(D)$ if and only if it is a simple semisimple ring (this may sound strange, but not every simple ring is semisimple; the terminology becomes a bit awkward here). Moreover, the uniqueness part of Corollary 3.58 also holds in this context.

The preceding paragraph outlines the so-called **Wedderburn-Artin theory**, developed by E. Artin some 20 years after Wedderburn's seminal work on finite dimensional algebras. Treating semisimple rings is not substantially harder than treating semisimple finite dimensional algebras. Nevertheless, we have decided to stick with the classical finite dimensional context, which is probably intuitively more clear and therefore easier to understand for a beginner. The interested reader will find a detailed exposition of the Wedderburn-Artin theory in a number of textbooks, in particular in [Bea99, FD93, Her68, Lam01, Pie82, Row91, Row08]. One of the ingredients of this theory are the *artinian rings*, which will be briefly touched upon in the next section.

3.10 Chain Conditions

The following three statements are equivalent for a vector space V: (i) V is infinite dimensional, (ii) there exists an infinite descending chain $V_1 \supsetneq V_2 \supsetneq \ldots$ of subspaces of V, (iii) there exists an infinite ascending chain $W_1 \subsetneq W_2 \subsetneq \ldots$ of subspaces of V. Indeed, each of (ii) and (iii) obviously implies (i), and the converses follow by taking a linearly independent subset $\{v_1, v_2, \ldots\}$ of V and defining V_n to be the linear span of $\{v_n, v_{n+1}, \ldots\}$ and W_n to be the linear span of $\{v_1, \ldots, v_n\}$. The following two notions can be therefore viewed as generalizations of finite dimensionality.

Definition 3.61 A module M is said to be **artinian** (or is said to satisfy the **descending chain condition** on submodules) if for every chain $M_1 \supseteq M_2 \supseteq \ldots$ of submodules of M, there exists $m \in \mathbb{N}$ such that $M_m = M_{m+1} = \ldots$.

Definition 3.62 A module N is said to be **noetherian** (or is said to satisfy the **ascending chain condition** on submodules) if for every chain $N_1 \subseteq N_2 \subseteq \ldots$ of submodules of N, there exists $n \in \mathbb{N}$ such that $N_n = N_{n+1} = \ldots$.

Let X be a set and $\mathscr{S}$ be a set of subsets of X. We say that $A \in \mathscr{S}$ is a **minimal element** in $\mathscr{S}$ (with respect to set inclusion) if there does not exist $Y \in \mathscr{S}$ such that $A \supsetneq Y$. Similarly, we say that B is a **maximal element** in $\mathscr{S}$ if there does not exist $Z \in \mathscr{S}$ such that $B \subsetneq Z$.

Lemma 3.63 *The following statements are equivalent for a module M:*

(i) *M is artinian.*
(ii) *Every nonempty set of submodules of M has a minimal element.*

Proof (i) $\Longrightarrow$ (ii). Let $\mathscr{S}$ be a nonempty set of submodules of M. Pick $M_1 \in \mathscr{S}$. If M_1 is not minimal, then there exists $M_2 \in \mathscr{S}$ such that $M_1 \supsetneq M_2$. If M_2 is not minimal, then $M_2 \supsetneq M_3$ for some $M_3 \in \mathscr{S}$. Continuing this process we arrive in a finite number of steps at a minimal element in $\mathscr{S}$; for if not, there would exist an infinite chain $M_1 \supsetneq M_2 \supsetneq \ldots$. (Note that the axiom of choice was used in this argument.)

(ii) $\Longrightarrow$ (i). Let $M_1 \supseteq M_2 \supseteq \ldots$ be a descending chain of submodules of M. If M_m is a minimal element of the set $\{M_1, M_2, \ldots\}$, then $M_m = M_{m+1} = \ldots$. $\square$

Lemma 3.64 *The following statements are equivalent for a module N:*

(i)   *N is noetherian.*
(ii) *Every nonempty set of submodules of N has a maximal element.*
(iii) *Every submodule of N is finitely generated.*

Proof (i) $\iff$ (ii). This can be proved in the same manner as Lemma 3.63, just the inclusions must be reversed.

(ii) $\implies$ (iii). Let P be a submodule of N. Denote by $\mathscr{S}$ the set of all finitely generated submodules of P. Since it contains 0, $\mathscr{S}$ is nonempty. Thus there exists a maximal element B in $\mathscr{S}$. We have to show that $B = P$. Let $b_1, \dots, b_r$ be generators of B. Take an arbitrary $p \in P$. The submodule generated by $p, b_1, \dots, b_r$ lies in $\mathscr{S}$ and contains B, and therefore it must be equal to B. In particular, $p \in B$.

(iii) $\implies$ (i). Let $N_1 \subseteq N_2 \subseteq \dots$ be a chain of submodules of N. Note that $\bigcup_{i=1}^{\infty} N_i$ is also a submodule of N, and is therefore generated by finitely many elements $c_1, \dots, c_q$. Each c_i lies in one of the N_j's, so there exists $n \in \mathbb{N}$ such that N_n contains all $c_1, \dots, c_q$. Hence $N_n = \bigcup_{i=1}^{\infty} N_i$, i.e., $N_n = N_{n+1} = \dots$. $\qquad\square$

Definition 3.65 A ring R is said to be **left (resp. right) artinian** if R is an artinian left (resp. right) R-module.

Definition 3.66 A ring R is said to be **left (resp. right) noetherian** if R is a noetherian left (resp. right) R-module.

A left artinian (resp. noetherian) ring is not necessarily right artinian (resp. noetherian), and vice versa. For simplicity we will consider only left artinian and left noetherian rings in this section. Of course, all statements that we will record have analogous right versions. If R is both left and right artinian (resp. noetherian), then we say that R is **artinian (resp. noetherian)**. A left artinian (resp. noetherian) commutative ring is obviously automatically artinian (resp. noetherian).

Since submodules of the left R-module R are the left ideals of R, R is left artinian if and only if for every chain $L_1 \supseteq L_2 \supseteq \dots$ of left ideals of R there exists $m \in \mathbb{N}$ such that $L_m = L_{m+1} = \dots$. Analogously, R is left noetherian if it satisfies the ascending chain condition on left ideals.

There is a non-obvious relation between these two classes of rings: A left artinian unital ring is left noetherian. But we will not prove this. On the other hand, the ring $\mathbb{Z}$ is noetherian but not artinian; this follows easily from the fact that the nonzero ideals of $\mathbb{Z}$ are $n\mathbb{Z}$, $n \in \mathbb{N}$, and that $n\mathbb{Z} \subseteq m\mathbb{Z}$ if and only if m divides n.

The artinian property turns out to be a perfect ring-theoretic substitute for finite dimensionality. The basic structure theory of finite dimensional algebras can be, sometimes with literally the same proofs and sometimes with an extra effort, generalized to artinian rings. But we will not go into details on that. Let us mention only the fundamental result, often referred in the literature as the Wedderburn-Artin theorem, which states that a ring R is simple and left artinian if and only if $R \cong M_n(D)$ where $n \in \mathbb{N}$ and D is a division ring. In Chap. 5 we will prove a version of the (more substantial) "only if" part, saying that a simple unital ring with a minimal left ideal is isomorphic to $M_n(D)$ (Corollary 5.34). Left artinian rings of course do have minimal left ideals by Lemma 3.63 (the set of all nonzero left ideals has a minimal element).

The noetherian property gives rise to a considerably richer class of rings. Lemma 3.64 implies that a ring R is left noetherian if and only if every left ideal of R is finitely generated. In particular, if every left ideal of R is *principal*, i.e., it is generated by a single element, then R is left noetherian. Besides $\mathbb{Z}$, the basic example of such a ring is $F[\omega]$ where F is a field. We assume that the reader has learned this fact in a basic algebra course. A deeper result, known as **Hilbert's basis theorem**, says that if T is a left noetherian unital ring, then so is $T[\omega]$. We will now prove a slightly more general version which is not limited to polynomials. However, the reader will notice that the polynomials are imitated in the proof.

Theorem 3.67 *Let R be a unital ring, T be a subring of R that contains the unity of R, and $d \in R$. If T is left noetherian, $T + dT = T + Td$, and R is generated by the set $T \cup \{d\}$, then R is left noetherian.*

Proof From $Td \subseteq T + dT$ it follows easily that $Td^s \subseteq T + dT + \cdots + d^sT$ for every $s \geq 1$. Similarly, $dT \subseteq T + Td$ yields $d^sT \subseteq T + Td + \cdots + Td^s$. Consequently,

$$T + dT + \cdots + d^sT = T + Td + \cdots + Td^s \text{ for every } s \geq 1. \qquad (3.4)$$

We denote the set in (3.4) by R_s; we also set $R_0 := T$. Clearly, $TR_s = R_sT = R_s$ for every $s \geq 0$.

Take a left ideal L of R. Our goal is to prove that L is finitely generated. Set $W_0 := L \cap T$ and

$$W_k := \{w \in T \mid d^kw \in R_{k-1} + L\}, \quad k \geq 1.$$

Using (3.4) one easily verifies that W_k is a left ideal of T. Since T is left noetherian, W_k is finitely generated: $W_k = \sum_{j=1}^{r_k} Tw_{jk}$ for some $w_{jk} \in W_k$. Next, using $dL \subseteq L$ and $dR_{k-1} \subseteq R_k$ we infer that $W_k \subseteq W_{k+1}$ for every $k \geq 0$. Therefore $W_n = W_{n+1} = \cdots$ for some $n \geq 0$.

By definition of W_k, $k \geq 1$, for each w_{jk} there is $\ell_{jk} \in L$ such that $\ell_{jk} - d^kw_{jk} \in R_{k-1}$; for $k = 0$ we simply take $\ell_{j0} = w_{j0} \in L$. Let $M := \sum_{k=0}^{n} \sum_{j=1}^{r_k} R\ell_{jk}$. We want to show that L is equal to M and is therefore finitely generated. Obviously $M \subseteq L$. Since R is generated by $T \cup \{d\}$ it can be deduced from (3.4) that $R = \bigcup_{s=0}^{\infty} R_s$. Therefore it suffices to show that $L \cap R_s \subseteq M$ for every $s \geq 0$. We proceed by induction on s. For $s = 0$ we have $L \cap R_0 = W_0 \subseteq M$, so let $s \geq 1$. Take $\ell \in L \cap R_s$ and let $w_s \in T$ be such that $\ell - d^sw_s \in R_{s-1}$.

First assume $s \leq n$. Since, by the very definition, $w_s \in W_s$, we have $w_s = \sum_{j=1}^{r_s} t_jw_{js}$ for some $t_j \in T$. By (3.4) there exist elements $t'_j \in T$ such that $d^st_j - t'_jd^s \in R_{s-1}$. A short calculation, based on $\ell_{js} - d^sw_{js} \in R_{s-1}$, shows that $\ell - \sum_{j=1}^{r_s} t'_j\ell_{js} \in R_{s-1}$. Since this is an element from L, we may use the induction assumption to conclude that it lies in M. As $\sum_{j=1}^{r_s} t'_j\ell_{js} \in M$ by definition, it follows that $\ell \in M$.

If $s > n$, then $w_s \in W_s = W_n$. As in the previous paragraph we can find $t_j, t'_j \in T$ such that $w_s = \sum_{j=1}^{r_n} t_jw_{jn}$ and $d^nt_j - t'_jd^n \in R_{n-1}$. One can verify that $\ell - \sum_{j=1}^{r_n} d^{s-n}t'_j\ell_{jn} \in R_{s-1}$. Therefore this element lies in M by induction assumption, yielding $\ell \in M$. $\square$

Example 3.68 The Weyl algebra $\mathscr{A}_1$ satisfies the conditions of Theorem 3.67 with $\mathscr{L}$ playing the role of T and D playing the role of d. Indeed, $\mathscr{L}$ is easily seen to be isomorphic to $F[\omega]$ and is therefore (both left and right) noetherian. From (1.7) it follows that $\mathscr{L} + D\mathscr{L} \subseteq \mathscr{L} + \mathscr{L}D$. Analogously we see that $\mathscr{L} + \mathscr{L}D \subseteq \mathscr{L} + D\mathscr{L}$. Of course, $\mathscr{A}_1$ is generated by $\mathscr{L} \cup \{D\}$. Therefore $\mathscr{A}_1$ is noetherian.

Exercises

3.1. Let R be a unital ring and M be a unital R-module. Show that M is cyclic if and only if $M \cong R/L$ for some left ideal L of R.

3.2. Let L and N be submodules of a module M. Show that:

   (a) $N/(N \cap L) \cong (N + L)/L$.
   (b) If $L \subseteq N$, then N/L is a submodule of M/L, and $(M/L)/(N/L) \cong M/N$.

   Remark: These two results, together with the standard fact that $M/\ker \varphi \cong N$ if $\varphi : M \to N$ is a surjective homomorphism, are often referred to as the **isomorphism theorems**. They are usually attributed to E. Noether. Analogous theorems, with analogous proofs, hold for other algebraic structures, in particular for groups and rings. The ring-theoretic versions of (a) and (b) read as follows: If I and J are ideals of a ring R, then $J/J \cap I \cong (J + I)/I$, and $(R/I)/(J/I) \cong R/J$ if $I \subseteq J$.

3.3. Let L be a submodule of a module M. Show that every submodule of M/L is of the form N/L, where N is a submodule of M which contains L. Similarly, if J is an ideal of a ring R, then every ideal of R/I is of the form J/I, where J is an ideal of R which contains I.

3.4. Let e and f be idempotents in a ring R. Show that the left ideals Re and Rf are isomorphic as R-modules if and only if there exist $u, v \in R$ such that $e = uv$ and $f = vu$. Moreover, if $R = M_n(F)$ with F a field this is further equivalent to the condition that e and f are similar matrices, i.e., there exists an invertible $t \in R$ such that $f = tet^{-1}$.

3.5. Let R be a unital ring, and let M be a free R-module with basis B. Note that every function from B into a unital R-module N can be (uniquely) extended to a homomorphism from M into N. Hence deduce that every unital R-module is the homomorphic image of a free R-module.

3.6. Is a submodule of a free module free?

3.7. A module P is said to be **projective** if for every surjective homomorphism $f : M \to N$ and every homomorphism $g : P \to N$ there exists a homomorphism $h : P \to M$ such that $fh = g$. Show that every free module is projective.

3.8. Note that a ring R is isomorphic to R° if and only if R has an antiautomorphism. Show that if R is such a ring, then so is $M_n(R)$. Find an example of a ring without antiautomorphisms.

   Hint: Where does an antiautomorphism send a left unity?

3.9. Let V be a vector space over a division ring D. Suppose V has an infinite countable basis. Show that the ring $\mathrm{End}_D(V)$ is isomorphic to the ring of all column finite $\mathbb{N} \times \mathbb{N}$ matrices over D^0, i.e., matrices with the property that each of their columns has only finitely many nonzero entries.

Remark: The column finite matrices indeed form a ring under the standard matrix operations. So do, for instance, the row finite matrices. On the other hand, the set of all $\mathbb{N} \times \mathbb{N}$ matrices is not a ring since one faces infinite summations when performing multiplication. (This set is thus merely an additive group.)

3.10. Describe the center of the ring $\mathrm{End}_D(V)$.

3.11. Let V be a finite dimensional vector space over a division ring D. We define rank(f), the **rank** of $f \in \mathrm{End}_D(V)$, as the dimension of im$f$. Show that rank$(f) = \dim_D V - \dim_D \ker f$. Further, show that $d(f, g) := \mathrm{rank}(f - g)$ defines a metric on $\mathrm{End}_D(V)$.

3.12. Let D be a division ring and let $A = (a_{ij}) \in M_n(D)$. Which of the following conditions imply that A is invertible?

  (a) A is left invertible.
  (b) A is not a left zero-divisor.
  (c) $\sum_{\sigma \in S_n} \mathrm{sgn}(\sigma) a_{1\sigma(1)} \ldots a_{n\sigma(n)} \neq 0$.

3.13. Show that every simple module over a unital ring is unital.

3.14. Let V be a nonzero vector space over a division ring D, and let $R = \mathrm{End}_D(V)$. As in Example 3.38, consider V as a (simple) R-module. Show that $\mathrm{End}_R(V) \cong D$.

3.15. Let M be a simple module over a commutative unital ring R. Show that the division ring $\mathrm{End}_R(M)$ from Schur's lemma is a field isomorphic to the factor ring R/U where U is a maximal ideal of R.

3.16. Show that a homomorphic image of a semisimple module is semisimple.

3.17. Show that finite dimensional simple algebras A and B are isomorphic if (and only if) there exists $n \in \mathbb{N}$ such that the algebras $M_n(A)$ and $M_n(B)$ are isomorphic.

3.18. The **left annihilator** of a subset I of a ring R is the left ideal $\mathrm{lann}_R(I) := \{x \in R \mid xI = 0\}$, and the **right annihilator** of I is the right ideal $\mathrm{rann}_R(I) := \{x \in R \mid Ix = 0\}$. Show that if R is semiprime and I is an ideal, then $\mathrm{lann}_R(I) = \mathrm{rann}_R(I)$ is an ideal, and if, additionally, R is a finite dimensional algebra, then $\mathrm{lann}_R(\mathrm{lann}_R(I)) = I$. Note that this formula is not, in general, valid if I is merely a left ideal. However, a similar formula that also involves the right annihilator does hold in this case. Find it!

3.19. Let L be a submodule of a module M. Show that M is artinian (resp. noetherian) if and only if both L and M/L are artinian (resp. noetherian).

3.20. Show that an injective endomorphism of an artinian module is automatically surjective. Similarly, a surjective endomorphism of a noetherian module is injective.

3.21. Suppose a module M is both artinian and noetherian. Show that for every endomorphism φ of M there exists $n \in \mathbb{N}$ such that $M = \operatorname{im} \varphi^n \oplus \ker \varphi^n$.

Remark: This result is known as **Fitting's lemma**.

3.22. Show that the polynomial ring $F[\Omega]$, where F is a field, is noetherian if and only if the set Ω is finite. Is the ring of continuous functions $C[a, b]$ noetherian?

3.23. Let V be a vector space over a division ring D. Show that the following statements are equivalent:

  (i) The space V is finite dimensional.
  (ii) The ring $\operatorname{End}_D(V)$ is left artinian.
  (iii) The ring $\operatorname{End}_D(V)$ is left noetherian.

Hint: The ring $M_n(D^o)$ has the natural structure of a vector space over D^o of dimension n^2, and its left ideals are automatically subspaces.

Chapter 4
Tensor Products

Sometimes we find some mathematical concept difficult to comprehend at the beginning. Later, when we get used to it, it appears so natural to us that we do not understand anymore what appeared so mysterious about it before. Perhaps the tensor product is a concept with which many have such an experience.

Our intention is to give a gentle introduction to tensor products. After a preliminary, unavoidable consideration of tensor products of vector spaces, we will proceed with tensor products of algebras. This will in particular yield a deeper insight into central simple algebras. In the final part of the chapter we will prove two classical results, the Double Centralizer Theorem and the Skolem-Noether Theorem, and thereby illustrate the usefulness and the strength of the approach based on tensor products and modules.

4.1 Concept of a Tensor Product

Take two vector spaces U and V over a field F. For simplicity assume, temporarily, that they are finite dimensional, $\dim_F U = m$ and $\dim_F V = n$. The tensor product of U and V is a vector space of dimension mn. One is usually interested in algebraic objects only up to isomorphism, and so might consider this description of the tensor product as satisfactory. However, it does not really give a clue about the idea upon which this concept is based. The tensor product serves as a device transforming bilinear maps into linear ones. This ambiguous description in some sense better reflects the essence of this concept. Before starting with a rigorous treatment, let us give an indication in what way the latter could be connected with an mn-dimensional space. Pick a basis $\{e_1, \ldots, e_m\}$ of U and a basis $\{f_1, \ldots, f_n\}$ of V. One easily notices that a *bilinear* map β from $U \times V$ into another vector space W is uniquely determined by its mn values $\beta(e_i, f_j)$, $1 \le i \le m$, $1 \le j \le n$. Now, take any vector space of dimension mn, choose its basis, and define a *linear* map $\overline{\beta}$ from this space into W so that the elements from the chosen basis are mapped into the elements $\beta(e_i, f_j)$. It is clear that $\overline{\beta}$ together with the chosen bases contains all information about β.

© Springer International Publishing Switzerland 2014   79
M. Brešar, *Introduction to Noncommutative Algebra*, Universitext,
DOI 10.1007/978-3-319-08693-4_4

Thus, instead of considering a bilinear map on $U \times V$, we can consider a linear map defined on a space of dimension $\dim_F U \cdot \dim_F V$.

We now proceed to the formal definition. Its major advantage comparing to the above discussion is that it is independent of basis. Before stating the definition we make a few comments in order to avoid a possible misapprehension.

Given any set X we can construct a vector space over F having X as a basis (cf. Sect. 2.5). Its elements are formal linear combinations of elements from X, i.e., they can be written as $\sum_{x \in X} \lambda_x x$ where all but finitely many $\lambda_x \in F$ are zero. Thus X basically plays the role of an index set. Operations in this space are defined in the obvious way. Ignoring the terms with zero coefficient we get another interpretation: For every element w in the vector space with basis X there exist a finite subset $\{x_1, \ldots, x_n\}$ of X and unique $\lambda_1, \ldots, \lambda_n \in F$ such that $w = \lambda_1 x_1 + \cdots + \lambda_n x_n$. We will deal with the situation where X is the Cartesian product $U \times V$ of spaces U and V. Let us emphasize that here we consider $U \times V$ only as a set, not as a vector space. Therefore, for example, $(0, 0)$ is just an element from the standard basis of the space with basis $U \times V$, it is not its zero element. Further, $(u + u', v)$ is an element from this basis and is therefore different from $(u, v) + (u', v)$ which is the sum of two basis elements, while $\lambda(u, v)$ coincides with $(\lambda u, v)$ only when $\lambda = 1$. We may thus think of the space with basis $U \times V$ as being "huge", incomparably larger than U and V.

Let us now state the definition. It concerns arbitrary vector spaces, not necessarily finite dimensional ones.

Definition 4.1 Let U and V be vector spaces over F. Denote by $\mathscr{Y}$ the vector space with basis $U \times V$. Let $\mathscr{N}$ be the subspace of $\mathscr{Y}$ generated by all elements of the form

$$(\lambda u + \lambda' u', v) - \lambda(u, v) - \lambda'(u', v),$$

$$(u, \lambda v + \lambda' v') - \lambda(u, v) - \lambda'(u, v'),$$

$\lambda, \lambda' \in F, u, u' \in U, v, v' \in V$. The **tensor product** of U and V is the factor vector space $\mathscr{Y}/\mathscr{N}$. It is denoted by $U \otimes V$ (or by $U \otimes_F V$ when we wish to emphasize that we consider spaces over F).

This definition is of little practical use. The characteristic property of $U \otimes V$, described in the following theorem, is the one that matters.

Theorem 4.2 *Let U and V be vector spaces over F. Then there exists a bilinear map $U \times V \to U \otimes V$, $(u, v) \mapsto u \otimes v$, such that:*

(a) *Every element in $U \otimes V$ is a sum of elements of the form $u \otimes v$, $u \in U$, $v \in V$.*
(b) *Given a bilinear map $\beta : U \times V \to W$, where W is a vector space over F, there exists a linear map $\overline{\beta} : U \otimes V \to W$ such that $\overline{\beta}(u \otimes v) = \beta(u, v)$ for all $u \in U, v \in V$.*

Moreover, properties (a) *and* (b) *characterize $U \otimes V$ up to isomorphism.*

Proof We define $u \otimes v$ as the coset $(u, v) + \mathcal{N}$. Considering the generators of $\mathcal{N}$ we see that the map $(u, v) \mapsto u \otimes v$ is bilinear. By definition, every element in $U \otimes V$ can be written as a linear combination of elements of the form $u \otimes v$. Since $\lambda(u \otimes v) = (\lambda u) \otimes v$, (a) follows.

Let $\beta : U \times V \to W$ be a bilinear map. Since $U \times V$ is a basis of $\mathcal{Y}$, there exists a linear map $B : \mathcal{Y} \to W$ such that $B\big((u, v)\big) = \beta(u, v)$ for all $u \in U, v \in V$ (cf. Remark 3.29). The bilinearity of β implies that $\mathcal{N} \subseteq \ker B$. Accordingly, we can define a linear map $\overline{\beta}$ on $\mathcal{Y}/\mathcal{N} = U \otimes V$ by $\overline{\beta}(x + \mathcal{N}) = B(x)$. In particular, $\overline{\beta}(u \otimes v) = \beta(u, v)$.

To prove the last statement, assume that T is another space for which there exists a bilinear map $U \times V \to T$, $(u, v) \mapsto u \odot v$, with properties (a) and (b). Since $(u, v) \mapsto u \otimes v$ is a bilinear map, there exists a linear map $\varphi : T \to U \otimes V$ satisfying $\varphi(u \odot v) = u \otimes v$. Analogously, there exists a linear map $\psi : U \otimes V \to T$ satisfying $\psi(u \otimes v) = u \odot v$. Consequently, φ and ψ are inverses of each other, and hence T and $U \otimes V$ are isomorphic vector spaces. □

Remark 4.3 From (a) it follows that $\overline{\beta}$ is a *unique* linear map from $U \otimes V$ into W sending $u \otimes v$ into $\beta(u, v)$.

Bilinear maps appear so often in mathematics that we simply cannot avoid them. However, they may be a bit clumsy to deal with. Theorem 4.2 shows that we can "glue" spaces U and V into the space $U \otimes V$ so that a bilinear map defined on $U \times V$ can be credibly represented by a linear map defined on $U \otimes V$. Linear maps are not only simpler than bilinear ones, they are the "proper" maps (i.e., the homomorphisms) in the context of vector spaces.

Example 4.4 Let A be an F-algebra. The multiplication on A is by definition a bilinear map $(x, y) \mapsto xy$ from $A \times A$ into A. Sometimes it is more appropriate to consider multiplication as a linear map from $A \otimes A$ into A, determined by $x \otimes y \mapsto xy$.

Let us mention that one can also define tensor products of modules. For modules over commutative rings this can be done in basically the same way as for vector spaces, the differences are merely terminological. More substantial changes are needed if rings are noncommutative.

4.2 Basic Properties of Tensor Products

The results of this and of the next couple of sections should convince the reader that the concept of the tensor product is natural. It is compatible with fundamental linear algebra concepts.

Let us first summarize the most relevant properties of the tensor product established so far. The tensor product of vector spaces U and V is a vector space $U \otimes V$ consisting of elements that can be written as

$$u_1 \otimes v_1 + u_2 \otimes v_2 + \cdots + u_n \otimes v_n, \quad u_i \in U, v_i \in V.$$

Such an expression is far from unique. After all, we have to take into account that the map $(u, v) \mapsto u \otimes v$ is bilinear, meaning that

$$(\lambda u + \lambda' u') \otimes v = \lambda(u \otimes v) + \lambda'(u' \otimes v),$$

$$u \otimes (\lambda v + \lambda' v') = \lambda(u \otimes v) + \lambda'(u \otimes v')$$

for all $\lambda, \lambda' \in F, u, u' \in U, v, v' \in V$. These formulas readily imply

$$u \otimes 0 = 0, \quad 0 \otimes v = 0$$

for all $u \in U, v \in V$. Last but not least, we have the so-called **universal property** (b) from Theorem 4.2, which might be easier to memorize through a commutative diagram:

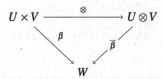

Here, $\otimes$ denotes the canonical map $(u, v) \mapsto u \otimes v$. Let us emphasize that $\otimes$ and β are *bilinear* maps, while $\overline{\beta}$ is *linear*.

Elements of the form $u \otimes v$ are called **simple tensors**. One should keep in mind that these are only generators of the space $U \otimes V$, not its typical elements. This warning may seem redundant at this point, but actually its disregard is a common source of errors. Standard tensor product considerations deal only with simple tensors, which may be misleading. Say, Theorem 4.2 describes $\overline{\beta}$ by $\overline{\beta}(u \otimes v) = \beta(u, v)$; this is all right since $\overline{\beta}$ is linear and so its action on $U \otimes V$ is uniquely determined by this rule. There are, however, other instances where an information about simple tensors is not sufficient.

Let us examine further properties of tensor products. First a word about notation: throughout this and the next section, U, U', V, V', W denote arbitrary vector spaces over the field F. The next lemma introduces the **tensor product of linear maps**.

Lemma 4.5 *If $\varphi : U \to U'$ and $\psi : V \to V'$ are linear maps, then there exists a unique linear map $U \otimes V \to U' \otimes V'$, which we denote by $\varphi \otimes \psi$, such that*

$$(\varphi \otimes \psi)(u \otimes v) = \varphi(u) \otimes \psi(v)$$

for all $u \in U, v \in V$.

Proof The linearity of φ and ψ, together with the bilinearity of $\otimes$, clearly imply that the map $(u, v) \mapsto \varphi(u) \otimes \psi(v)$ is bilinear. Theorem 4.2 thus shows that the linear map sending $u \otimes v$ into $\varphi(u) \otimes \psi(v)$ indeed exists. Its uniqueness is obvious. □

The lemma basically states that the map given by

$$\sum_i u_i \otimes v_i \mapsto \sum_i \varphi(u_i) \otimes \psi(v_i)$$

is well-defined. The given proof is quite illustrative. In order to establish something about tensor products one usually begins by introducing a bilinear map, and then, by the universal property, transforms it into a linear one. As a rule this quickly yields conclusions which may not be so easily obtained by other means.

It is clear that for linear maps with appropriate domains and codomains we have

$$(\varphi \otimes \psi)(\varphi' \otimes \psi') = \varphi\varphi' \otimes \psi\psi'.$$

Consequently, if φ and ψ are isomorphisms, then so is $\varphi \otimes \psi$, and

$$(\varphi \otimes \psi)^{-1} = \varphi^{-1} \otimes \psi^{-1}.$$

This in particular shows the following:

$$U \cong U' \text{ and } V \cong V' \Longrightarrow U \otimes V \cong U' \otimes V'.$$

Let us also mention the formulas

$$(\lambda\varphi + \lambda'\varphi') \otimes \psi = \lambda(\varphi \otimes \psi) + \lambda'(\varphi' \otimes \psi),$$

$$\varphi \otimes (\lambda\psi + \lambda'\psi') = \lambda(\varphi \otimes \psi) + \lambda'(\varphi \otimes \psi'),$$

which follow immediately from the definition.

In the next lemmas we prove that the tensor product is commutative and associative. One should not understand this literally; such statements can only be true up to isomorphism. However, such abuse of terminology is quite standard.

Lemma 4.6 $U \otimes V \cong V \otimes U$.

Proof The map $U \times V \to V \otimes U$, $(u, v) \mapsto v \otimes u$, is bilinear, hence there exists a linear map $U \otimes V \to V \otimes U$, $u \otimes v \mapsto v \otimes u$. Similarly there is a linear map $V \otimes U \to U \otimes V$, $v \otimes u \mapsto u \otimes v$. Obviously, these two maps are inverses of each other. $\square$

Lemma 4.7 $(U \otimes V) \otimes W \cong U \otimes (V \otimes W)$.

Proof The bulk of the proof is to show that the rule $(u \otimes v) \otimes w \mapsto u \otimes (v \otimes w)$ determines a well-defined map from $(U \otimes V) \otimes W$ to $U \otimes (V \otimes W)$. Since the elements $(u \otimes v) \otimes w$ generate the vector space $(U \otimes V) \otimes W$ (why?), the symmetrically defined map $u \otimes (v \otimes w) \mapsto (u \otimes v) \otimes w$ is then obviously its inverse.

Take $w \in W$. The map $U \times V \to U \otimes (V \otimes W)$, $(u, v) \mapsto u \otimes (v \otimes w)$, is bilinear, and hence $u \otimes v \mapsto u \otimes (v \otimes w)$ determines a well-defined map from $U \otimes V$ to $U \otimes (V \otimes W)$. This is true for every $w \in W$. Accordingly,

$$\left(\sum_i u_i \otimes v_i, w \right) \mapsto \sum_i u_i \otimes (v_i \otimes w)$$

is a well-defined bilinear map from $(U \otimes V) \times W$ to $U \otimes (V \otimes W)$. Therefore the map determined by $(u \otimes v) \otimes w \mapsto u \otimes (v \otimes w)$ is indeed well-defined. $\square$

One usually identifies $(U \otimes V) \otimes W$ and $U \otimes (V \otimes W)$, and denotes each of these two spaces simply by

$$U \otimes V \otimes W.$$

Alternatively one can introduce $U \otimes V \otimes W$ via trilinear maps; the resulting space is the same up to isomorphism. Of course, the tensor product of more than three spaces can also be defined.

4.3 Linear (In)dependence in Tensor Products

An element in $U \otimes V$ can be written as a sum of simple tensors in many different ways. Fortunately, this non-uniqueness has its limitations.

Lemma 4.8 *Let $e_1, \ldots, e_n \in U$ be linearly independent. If $v_1, \ldots, v_n \in V$ are such that*

$$e_1 \otimes v_1 + \cdots + e_n \otimes v_n = 0,$$

then each $v_i = 0$.

Proof Let $f_i \colon U \to F$ be a linear functional such that $f_i(e_j) = \delta_{ij}$ for $j = 1, \ldots, n$ (cf. Theorem 3.28 and Remark 3.29). The map $(u, v) \mapsto f_i(u)v$ is bilinear, and so there exists a linear map $\alpha_i : U \otimes V \to V$ satisfying $\alpha_i(u \otimes v) = f_i(u)v$. Consequently,

$$v_i = \sum_{j=1}^{n} \alpha_i(e_j \otimes v_j) = \alpha_i \left(\sum_{j=1}^{n} e_j \otimes v_j \right) = 0.$$ $\square$

All results in this section will follow from this simple lemma. Let us first point out that in particular it shows that for all $u \in U$ and $v \in V$,

$$u \otimes v = 0 \implies u = 0 \text{ or } v = 0.$$

Incidentally, the tensor product of modules over rings does not always enjoy this property.

We continue with a useful technical improvement of Lemma 4.8.

Lemma 4.9 *Let $e_1, \ldots, e_n \in U$ be linearly independent. If $v_1, \ldots, v_n \in V$ are such that*

$$e_1 \otimes v_1 + \cdots + e_n \otimes v_n = w_1 \otimes z_1 + \cdots + w_m \otimes z_m$$

for some $w_1 \ldots, w_m \in U$ and $z_1, \ldots, z_m \in V$, then each v_i is a linear combination of $z_1, \ldots, z_m$.

Proof The set $\{e_1, \ldots, e_n\}$ can be extended to a basis of the linear span of the set $\{e_1, \ldots, e_n, w_1, \ldots, w_m\}$. Denote the additional vectors (if there are any) in this basis by $e_{n+1}, \ldots, e_p$. Write each w_j as a linear combination of basis vectors. Then, using the bilinearity of $\otimes$, we see that the given identity can be rewritten as

$$e_1 \otimes v_1' + \cdots + e_n \otimes v_n' + e_{n+1} \otimes z_{n+1}' + \cdots + e_p \otimes z_p' = 0,$$

where each v_i' is the sum of v_i and a linear combination of the z_j's. Lemma 4.8 tells us that $v_i' = 0$. □

Let us also record the obvious analogues of the above lemmas.

Lemma 4.10 *Let $f_1, \ldots, f_n \in V$ be linearly independent. If $u_1, \ldots, u_n \in U$ are such that*

$$u_1 \otimes f_1 + \cdots + u_n \otimes f_n = 0,$$

then each $u_i = 0$.

Lemma 4.11 *Let $f_1, \ldots, f_n \in V$ be linearly independent. If $u_1, \ldots, u_n \in U$ are such that*

$$u_1 \otimes f_1 + \cdots + u_n \otimes f_n = w_1 \otimes z_1 + \cdots + w_m \otimes z_m$$

for some $w_1, \ldots, w_m \in U$ and $z_1 \ldots, z_m \in V$, then each u_i is a linear combination of $w_1, \ldots, w_m$.

Bases of U and V give rise to bases of $U \otimes V$:

Theorem 4.12 *If $\{e_i \mid i \in I\}$ is a basis of U, and $\{f_j \mid j \in J\}$ is a basis of V, then $\{e_i \otimes f_j \mid i \in I, j \in J\}$ is a basis of $U \otimes V$.*

Proof Writing $u \in U$ as a linear combination of the e_i's, and $v \in V$ as a linear combination of the f_j's, it follows, by the bilinearity of $\otimes$, that $u \otimes v$ is a linear combination of the $e_i \otimes f_j$'s. Therefore the set $\{e_i \otimes f_j \mid i \in I, j \in J\}$ generates $U \otimes V$. It remains to show that it is linearly independent. By definition, a subset of a vector space is linearly independent if each of its finite subsets is linearly independent.

Every finite subset of $\{e_i \otimes f_j \mid i \in I, j \in J\}$ is contained in a set of the form $\{e_{i_k} \otimes f_{j_l} \mid k = 1, \dots, m, \ l = 1, \dots, n\}$. Therefore it suffices to show that such sets are linearly independent. Assume, therefore, that

$$\sum_{k=1}^{m} \sum_{l=1}^{n} \lambda_{kl} (e_{i_k} \otimes f_{j_l}) = 0$$

for some $\lambda_{kl} \in F$. We can rewrite this as

$$\sum_{k=1}^{m} e_{i_k} \otimes \left(\sum_{l=1}^{n} \lambda_{kl} f_{j_l} \right) = 0.$$

Since $e_{i_1}, \dots, e_{i_m}$ are linearly independent, $\sum_{l=1}^{n} \lambda_{kl} f_{j_l} = 0$ for all k by Lemma 4.8. The linear independence of $f_{j_1}, \dots, f_{j_n}$ yields $\lambda_{kl} = 0$ for all k and l.   □

The next two corollaries are immediate.

Corollary 4.13 *If $\{e_i \mid i \in I\}$ is a basis of U, then every element in $U \otimes V$ can be written uniquely in the form $\sum_{i \in I} e_i \otimes v_i$.*

If I is an infinite set, then this should be understood as that all but finitely many v_i's are zero.

Corollary 4.14 *If $\{f_j \mid j \in J\}$ is a basis of V, then every element in $U \otimes V$ can be written uniquely in the form $\sum_{j \in J} u_j \otimes f_j$.*

Remark 4.15 Another useful observation is that every nonzero element $x \in U \otimes V$ can be written as $x = \sum_{i=1}^{d} u_i \otimes v_i$ where $u_1, \dots, u_d$ are linearly independent vectors in U and $v_1, \dots, v_d$ are linearly independent vectors in V. The proof is easy: just require the minimality of d with respect to all possible expressions of x, and the independence of the u_i's and of the v_i's clearly follows.

Let now I and J be finite sets. Theorem 4.12 then yields the formula announced at the beginning of the chapter.

Corollary 4.16 *If U and V are finite dimensional vector spaces over F, then*

$$\dim_F U \otimes V = \dim_F U \cdot \dim_F V.$$

Observe the analogy with the formula on direct sums:

$$\dim_F U \oplus V = \dim_F U + \dim_F V.$$

Remark 4.17 Corollary 4.16 resembles another formula from basic linear algebra: The dimension of $\mathrm{Hom}_F(U, V)$, the vector space of all linear maps from U to V, is equal to $\dim_F U \cdot \dim_F V$. Therefore the spaces $\mathrm{Hom}_F(U, V)$ and $U \otimes V$ are isomorphic. What we wish to point out is a canonical isomorphism—yet not between exactly these two spaces. Let U^* be the dual space of U (i.e., the space of all linear functionals on U). Instead of $U \otimes V$ we consider $V \otimes U^*$ which is a space of the same dimension. Given $v \in V$ and $f \in U^*$, define $T_{v,f} \in \mathrm{Hom}_F(U, V)$ by $T_{v,f}(u) = f(u)v$. Note that $T_{v,f}$ has rank one provided that $v \neq 0$ and $f \neq 0$, and conversely, every linear map $U \to V$ of rank one is of the form $T_{v,f}$ for some $0 \neq v \in V$ and $0 \neq f \in U^*$. The map $(v, f) \mapsto T_{v,f}$ is bilinear, hence there exists a linear map

$$\varphi : V \otimes U^* \to \mathrm{Hom}_F(U, V), \quad \varphi(v \otimes f) = T_{v,f}.$$

As every nonzero map in $\mathrm{Hom}_F(U, V)$ is a sum of maps of rank one, φ is surjective. But then φ is an isomorphism for it is linear and the spaces $V \otimes U^*$ and $\mathrm{Hom}_F(U, V)$ have the same dimension. One often simply identifies $v \otimes f$ with $T_{v,f}$, and thus considers simple tensors $v \otimes f$ as rank one maps given by

$$(v \otimes f)(u) = f(u)v.$$

This is commonly used regardless of whether U and V are finite dimensional or not.

Remark 4.18 The following elementary facts

$$U \otimes F \cong U \quad \text{and} \quad F \otimes V \cong V$$

also deserve to be mentioned. The proofs are immediate. By bilinearity of $\otimes$ every element in $U \otimes F$ can be written as $u \otimes 1$. Therefore $u \otimes 1 \mapsto u$ is an isomorphism from $U \otimes F$ onto U. Similarly, $1 \otimes v \mapsto v$ is an isomorphism from $F \otimes V$ onto V.

4.4 Tensor Product of Algebras

After a linear algebra inset we are now ready to return to themes that are in the spirit of this book. Let A and B be algebras over a field F. The vector space $A \otimes B$ can be turned into an algebra by defining multiplication in a simple, natural way.

Lemma 4.19 *If A and B are F-algebras, then $A \otimes B$ is an F-algebra relative to the multiplication determined by*

$$(x \otimes y)(z \otimes w) = xz \otimes yw$$

for all $x, z \in A$, $y, w \in B$.

Proof Pick $z \in A$ and $w \in B$. Let R_z and R_w denote the corresponding right multiplication maps. By Lemma 4.5 we have the map $R_z \otimes R_w \in \mathrm{End}_F(A \otimes B)$ that sends $x \otimes y$ into $xz \otimes yw$. Since

$$A \times B \to \mathrm{End}_F(A \otimes B), \ (z, w) \mapsto R_z \otimes R_w,$$

is a bilinear map, there exists a linear map

$$\varphi : A \otimes B \to \mathrm{End}_F(A \otimes B), \ \varphi(z \otimes w) = R_z \otimes R_w.$$

We now define the product of $r, s \in A \otimes B$ as follows:

$$rs := \varphi(s)(r).$$

This multiplication is obviously bilinear, and for all $x, z \in A$, $y, w \in B$ we have

$$(x \otimes y)(z \otimes w) = \varphi(z \otimes w)(x \otimes y) = (R_z \otimes R_w)(x \otimes y) = xz \otimes yw.$$

Finally, from the associativity in A and B we immediately infer that

$$\Big((x \otimes y)(z \otimes w)\Big)(u \otimes v) = (x \otimes y)\Big((z \otimes w)(u \otimes v)\Big).$$

Since every element in $A \otimes B$ is a sum of simple tensors, this implies that our multiplication is associative. $\square$

Lemma 4.19 basically states that

$$\Big(\sum_i x_i \otimes y_i\Big)\Big(\sum_j z_j \otimes w_j\Big) := \sum_{i,j} x_i z_j \otimes y_i w_j$$

is a well-defined operation on $A \otimes B$. From now on we will always consider $A \otimes B$, the **tensor product of algebras** A and B, as an algebra in which multiplication is defined by the above formula.

By Lemma 4.6 we know that $A \otimes B$ and $B \otimes A$ are isomorphic as vector spaces. The canonical isomorphism $x \otimes y \mapsto y \otimes x$ also preserves multiplication. Therefore we have an algebra isomorphism

$$A \otimes B \cong B \otimes A.$$

Similarly, just as for tensor products of vector spaces we see that

$$(A \otimes B) \otimes C \cong A \otimes (B \otimes C),$$

$$A \otimes F \cong A \quad \text{and} \quad F \otimes B \cong B,$$

$$A \cong A' \text{ and } B \cong B' \implies A \otimes B \cong A' \otimes B'$$

holds for all F-algebras A, A', B, B', and C.

Let S be a subalgebra of A and T be a subalgebra of B. Regarding S and T as algebras we can form their tensor product $S \otimes T$, which is clearly canonically isomorphic to the subalgebra of $A \otimes B$ consisting of all elements of the form $\sum_i s_i \otimes t_i$, $s_i \in S, t_i \in T$. In this sense we can consider $S \otimes T$ as a subalgebra of $A \otimes B$.

We continue with illuminating examples.

Example 4.20 Let G and H be groups. What is the tensor product of group algebras $F[G]$ and $F[H]$? From Theorem 4.12 we know that the set $\{g \otimes h \mid g \in G, h \in H\}$ is a basis of $F[G] \otimes F[G]$. By definition, $G \times H$ is a basis of the group algebra $F[G \times H]$. Therefore there exists a bijective linear map $F[G] \otimes F[H] \to F[G \times H]$ determined by $g \otimes h \mapsto (g, h)$. This map is clearly also multiplicative, so we have

$$F[G] \otimes F[H] \cong F[G \times H].$$

Example 4.21 Let A be an arbitrary F-algebra. We claim that

$$A \otimes F[\omega] \cong A[\omega].$$

Indeed, using Corollary 4.14 we see that every element in $A \otimes F[\omega]$ can be *uniquely* written as $\sum_i a_i \otimes \omega^i$. It is now straightforward to check that the map

$$A \otimes F[\omega] \to A[\omega], \quad \sum_i a_i \otimes \omega^i \mapsto \sum_i a_i \omega^i,$$

is an algebra isomorphism.

In particular, $F[\omega_1] \otimes F[\omega_2] \cong (F[\omega_1])[\omega_2] \cong F[\omega_1, \omega_2]$. Note that, via this isomorphism, the element $f \otimes g \in F[\omega_1] \otimes F[\omega_2]$ corresponds to the polynomial $f(\omega_1)g(\omega_2) \in F[\omega_1, \omega_2]$. It is in fact common not to distinguish between the two, and thus consider $f \otimes g$ as the polynomial

$$(f \otimes g)(\omega_1, \omega_2) = f(\omega_1)g(\omega_2).$$

Example 4.22 For every F-algebra A and every $n \geq 1$ we have

$$M_n(F) \otimes A \cong M_n(A). \tag{4.1}$$

The proof is similar to that from the preceding example. After noticing that every element in $M_n(F) \otimes A$ can be uniquely written as the sum of simple tensors $E_{ij} \otimes a_{ij}$, where E_{ij} are the standard matrix units, it readily follows that

$$M_n(F) \otimes A \to M_n(A), \quad \sum_{i=1}^{n} \sum_{j=1}^{n} E_{ij} \otimes a_{ij} \mapsto \sum_{i=1}^{n} \sum_{j=1}^{n} a_{ij} E_{ij},$$

is an algebra isomorphism. This map should be considered as canonical—one often identifies $(\lambda_{ij}) \otimes a \in M_n(F) \otimes A$ with the matrix in $M_n(A)$ whose (i, j) entry is $\lambda_{ij}a$.

As an important special case of (4.1) we have

$$M_n(F) \otimes M_m(F) \cong M_{nm}(F). \tag{4.2}$$

Namely, $M_n(M_m(F)) \cong M_{nm}(F)$. This is basically just a rewording of the well-known fact that matrices can be added and multiplied blockwise. A detailed proof is perhaps a bit tedious, but straightforward.

Alternatively, we can state (4.2) as

$$\operatorname{End}_F(U) \otimes \operatorname{End}_F(V) \cong \operatorname{End}_F(U \otimes V),$$

where U and V are finite dimensional vector spaces over F.

Example 4.23 Let V be a vector space with a countably infinite basis $\{e_1, e_2, \ldots\}$. Set $A := \operatorname{End}_F(V)$ and define $f_{ij} \in A$, $i, j = 1, 2$, according to

$$f_{11}(e_{2k-1}) = e_{2k-1}, \quad f_{11}(e_{2k}) = 0,$$

$$f_{12}(e_{2k-1}) = 0, \quad f_{12}(e_{2k}) = e_{2k-1},$$

$$f_{21}(e_{2k-1}) = e_{2k}, \quad f_{21}(e_{2k}) = 0,$$

$$f_{22}(e_{2k-1}) = 0, \quad f_{22}(e_{2k}) = e_{2k}$$

for all $k \in \mathbb{N}$. One can check that $\{f_{11}, f_{12}, f_{21}, f_{22}\}$ forms a set of 2×2 matrix units of A. Therefore $A \cong M_2(f_{11}Af_{11})$ by Lemma 2.52 (and Remark 2.53). It is an easy exercise to show that $f_{11}Af_{11} \cong \operatorname{End}_F(V_1)$ where V_1 is the subspace of V generated by all e_{2k-1}, $k \in \mathbb{N}$. As the spaces V_1 and V are isomorphic, we actually have $\operatorname{End}_F(V_1) \cong A$, and hence $A \cong M_2(A)$. It is not difficult to generalize this to

$$A \cong M_n(A) \quad \text{for every } n \in \mathbb{N}.$$

We can now better understand and appreciate the uniqueness in Wedderburn's theorem (Theorem 3.58). Further, now we see that the tensor product of algebras does not obey the "cancellation law", i.e.,

$$S \otimes A \cong T \otimes A \not\Longrightarrow S \cong T.$$

Namely, by (4.1) we have $F \otimes A \cong M_n(F) \otimes A$ for every $n \in \mathbb{N}$.

4.5 Multiplication Algebra and Tensor Products

Let A and B be unital algebras. Then $1 \otimes 1$ is clearly the unity of $A \otimes B$. One usually identifies A with $A \otimes 1 := \{a \otimes 1 \mid a \in A\} (= A \otimes F)$ and B with $1 \otimes B := \{1 \otimes b \mid b \in B\}$, and in this way considers A and B as subalgebras of $A \otimes B$. Obviously, elements from A commute with elements from B, and the algebra $A \otimes B$ is generated by $A \cup B$. If we add the statement of Lemma 4.8 to these properties, we get a characterization of the tensor product of unital algebras:

Lemma 4.24 *Let T be a unital algebra. Suppose T contains subalgebras A and B which contain the unity of T and satisfy*

(a) $ab = ba$ *for all $a \in A$, $b \in B$.*
(b) T *is generated by $A \cup B$.*
(c) *If $a_1, \dots, a_n \in A$ are linearly independent and $b_1, \dots, b_n \in B$ are arbitrary, then $a_1 b_1 + \cdots + a_n b_n = 0$ implies $b_1 = \cdots = b_n = 0$.*

Then $A \otimes B \cong T$ under the map $a \otimes b \mapsto ab$.

Proof The standard universal property argument shows that there exists a linear map $\varphi : A \otimes B \to T$ such that $\varphi(a \otimes b) = ab$ for all $a \in A$, $b \in B$. According to (a), (b), and the assumption that A and B contain the unity of T, every element in T is of the form $a_1 b_1 + \cdots + a_n b_n$ with $a_i \in A$ and $b_i \in B$. Therefore φ is surjective. Since every $x \in A \otimes B$ can be written as $x = a_1 \otimes b_1 + \cdots + a_n \otimes b_n$, where $a_1, \dots, a_n$ are linearly independent (cf. Remark 4.15), we see from (c) that $\varphi(x) = 0$ implies $x = 0$. Therefore φ is injective. Finally, using (a) we obtain

$$\varphi\big((a \otimes b)(a' \otimes b')\big) = \varphi(aa' \otimes bb') = aa'bb' = aba'b' = \varphi(a \otimes b)\varphi(a' \otimes b'),$$

which shows that φ is an algebra isomorphism. $\square$

Remark 4.25 It is instructive to compare (a)–(c) with analogous properties of the direct product. Suppose now that an algebra S contains subalgebras A and B such that

(a') $ab = ba = 0$ for all $a \in A$, $b \in B$.
(b') S is the sum of A and B.
(c') If $a \in A$ and $b \in B$ are such that $a + b = 0$, then $a = b = 0$.

Then $A \times B \cong S$ under the map $(a, b) \mapsto a + b$.

As it will be apparent from the next proof, the situation considered in Lemma 4.24 was indirectly met already in Sect. 1.5 when studying the multiplication algebra $M(A)$ of a central simple algebra A.

Theorem 4.26 *If A is a central simple algebra, then $A \otimes A^\circ \cong M(A)$ under the map $a \otimes b \mapsto L_a R_b$.*

Proof The sets $\mathscr{L} = \{L_a \mid a \in A\}$ and $\mathscr{R} = \{R_b \mid b \in A\}$ of all left and right multiplication maps are subalgebras of $M(A)$. By definition, $M(A)$ is generated by $\mathscr{L} \cup \mathscr{R}$. Obviously, elements from $\mathscr{L}$ commute with elements from $\mathscr{R}$. It is clear that $a_1, \ldots, a_n \in A$ are linearly independent if and only if $L_{a_1}, \ldots, L_{a_n} \in \mathscr{L}$ are linearly independent. In this case $L_{a_1} R_{b_1} + \cdots + L_{a_n} R_{b_n} = 0$ implies $b_i = 0$ (and hence $R_{b_i} = 0$) by Lemma 1.24. Lemma 4.24 therefore tells us that

$$\mathscr{L} \otimes \mathscr{R} \cong M(A) \text{ under the map } L_a \otimes R_b \mapsto L_a R_b.$$

We can polish this a little bit. Note that $A \cong \mathscr{L}$ under $a \mapsto L_a$. Since $R_{bb'} = R_{b'} R_b$, there is no reason why A and $\mathscr{R}$ should be isomorphic. However, A° and $\mathscr{R}$ are; $b \mapsto R_b$ is, of course, an isomorphism. Making use of Lemma 4.5 (and the comments following this lemma) we thus see that

$$A \otimes A^{\circ} \cong \mathscr{L} \otimes \mathscr{R} \text{ under the map } a \otimes b \mapsto L_a \otimes R_b.$$

Composing the isomorphisms $a \otimes b \mapsto L_a \otimes R_b$ and $L_a \otimes R_b \mapsto L_a R_b$ we arrive at the desired conclusion. $\qquad\qquad\qquad\qquad\qquad\qquad\qquad\qquad\qquad\qquad\qquad\quad\square$

Lemma 1.24 can now be sharpened as follows.

Corollary 4.27 *If A is a central simple algebra, then for all $a_i, b_i \in A$ we have $\sum_{i=1}^{n} L_{a_i} R_{b_i} = 0$ if and only if $\sum_{i=1}^{n} a_i \otimes b_i = 0$.*

Corollary 4.28 *If A is a finite dimensional central simple algebra, then $A \otimes A^{\circ} \cong M_n(F)$, where $n = [A : F]$.*

Proof Apply Lemma 1.25, which says that $M(A) = \text{End}_F(A)$, together with the standard fact $\text{End}_F(A) \cong M_n(F)$. $\qquad\qquad\qquad\qquad\qquad\qquad\qquad\qquad\quad\square$

Example 4.29 The conjugation $h \mapsto \bar{h}$ is an antiautomorphism of $\mathbb{H}$. Therefore $\mathbb{H}^{\circ} \cong \mathbb{H}$ under the map $h \mapsto \bar{h}$, and hence $\mathbb{H} \otimes_{\mathbb{R}} \mathbb{H}^{\circ} \cong \mathbb{H} \otimes_{\mathbb{R}} \mathbb{H}$. As $\mathbb{H} \otimes_{\mathbb{R}} \mathbb{H}^{\circ} \cong M_4(\mathbb{R})$ by Corollary 4.28, we have

$$\mathbb{H} \otimes_{\mathbb{R}} \mathbb{H} \cong M_4(\mathbb{R}).$$

4.6 Centralizers in Tensor Products

The center $Z(A)$ of an algebra A is the set of elements in A that commute with *all* elements in A. We now extend this concept by considering sets of elements in A that commute with *some* elements in A.

Definition 4.30 Let S be a subset of an algebra A. The **centralizer of S in A** is the set

$$C_A(S) := \{a \in A \mid as = sa \text{ for every } s \in S\}.$$

In the same way one defines the centralizer of a subset of a ring, but we shall deal only with centralizers in algebras. Let us also mention that we have encountered centralizers of singletons in the proof of Wedderburn's theorem on finite division rings.

First we record a few simple remarks to build intuition. The centralizer of any set S is a subalgebra of A which contains $Z(A)$. On the other hand, the centralizer of S coincides with the centralizer of the subalgebra generated by S and $Z(A)$. Therefore one usually considers only centralizers of subalgebras containing the center. One might expect that, roughly speaking, a large subalgebra has a small centralizer, and a small subalgebra has a large centralizer. Let us support this inexact statement with the following obvious facts:

$$S_1 \subseteq S_2 \implies C_A(S_2) \subseteq C_A(S_1),$$

$$C_A(A) = Z(A) \quad \text{and} \quad C_A(Z(A)) = A.$$

Condition (a) in Lemma 4.24 can be read as $B \subseteq C_T(A)$ (or $A \subseteq C_T(B)$). This suggests that the concept of the centralizer plays a role in tensor products of algebras. Our starting point is

Proposition 4.31 *Let A and B be unital algebras. If S is a subalgebra of A that contains the unity of A, and T is a subalgebra of B that contains the unity of B, then*

$$C_{A \otimes B}(S \otimes T) = C_A(S) \otimes C_B(T).$$

Proof It is enough to prove $C_{A \otimes B}(S \otimes T) \subseteq C_A(S) \otimes C_B(T)$, since the converse inclusion is trivial. Let $c \in C_{A \otimes B}(S \otimes T)$. We can write $c = \sum_{i=1}^{d} y_i \otimes z_i$ where $y_1, \ldots, y_d \in A$ are linearly independent and $z_1, \ldots, z_d \in B$ are linearly independent (cf. Remark 4.15). From $c(s \otimes 1) = (s \otimes 1)c, s \in S$, it follows that

$$\sum_{i=1}^{d} (y_i s - s y_i) \otimes z_i = 0.$$

Lemma 4.10 implies that $y_i s - s y_i = 0$ for every i and every $s \in S$. Therefore each y_i lies in $C_A(S)$. Analogously, by considering $c(1 \otimes t) = (1 \otimes t)c, t \in T$, we infer that each z_i lies in $C_B(T)$. Thus $c \in C_A(S) \otimes C_B(T)$. $\quad\square$

Corollary 4.32 *If A and B are unital algebras, then $Z(A \otimes B) = Z(A) \otimes Z(B)$.*

Proof Use Proposition 4.31 for $S = A$ and $T = B$. $\quad\square$

In the statement of the next corollary we identify A with $A \otimes 1$ and B with $1 \otimes B$.

Corollary 4.33 *If A and B are central algebras, then $A \otimes B$ is also a central algebra, $C_{A \otimes B}(A) = B$ and $C_{A \otimes B}(B) = A$.*

Proof The first assertion follows from Corollary 4.32, the second one from Proposition 4.31 by taking $S = A$ and $T = F (= F \cdot 1)$, and the third one from Proposition 4.31 by taking $S = F$ and $T = B$. $\qquad\qquad\qquad\qquad\qquad\qquad \Box$

Corollary 4.34 *Let A be a unital algebra and let S be its subalgebra that contains the unity of A. If B is a central algebra, then $C_{A \otimes B}(S \otimes B) \cong C_A(S)$. In particular, $C_{M_n(A)}(M_n(S)) \cong C_A(S)$ for every $n \in \mathbb{N}$.*

Proof Proposition 4.31 shows that $C_{A \otimes B}(S \otimes B) = C_A(S) \otimes F \cong C_A(S)$. The last assertion follows by taking $M_n(F)$ for B (cf. Example 4.22). $\qquad\qquad\qquad \Box$

If S is a subalgebra of A, then we can consider A as a (left) S-module (cf. Example 3.8). Therefore it makes sense to speak about $\mathrm{End}_S(A)$.

Corollary 4.35 *Let A be a finite dimensional central simple algebra. If S is a subalgebra of A that contains the unity of A, then $C_A(S) \otimes A^\circ \cong \mathrm{End}_S(A)$.*

Proof Theorem 4.26, along with Lemma 1.25, tells us that $a \otimes b \mapsto L_a R_b$ defines an isomorphism from $A \otimes A^\circ$ onto $\mathrm{End}_F(A)$. Accordingly, the centralizer of $S \otimes F$ in $A \otimes A^\circ$ is isomorphic to the centralizer of $\{L_s \mid s \in S\}$ in $\mathrm{End}_F(A)$. The former equals $C_A(S) \otimes A^\circ$ by Proposition 4.31, while the latter is nothing but $\mathrm{End}_S(A)$. $\Box$

4.7 Scalar Extension (The "Right" Approach)

We are now in a position to place the notion of the scalar extension, already discussed in Sect. 2.11, on a solid foundation. Recall the basic idea behind the definition: Given an F-algebra A and an extension field K of F, we wish to construct a K-algebra A_K that has essentially the same multiplication as A. As we have seen, relying on a fixed chosen basis this can be achieved by elementary means. We will now show that by making use of the tensor product this construction can be accomplished independently of basis.

Recall that we can consider an extension field K of F as an F-algebra. Therefore $K \otimes_F A$ is an F-algebra. The point is to convert it into a K-algebra.

Lemma 4.36 *Let A be an F-algebra and let K be an extension field of F. Then $K \otimes_F A$ is a K-algebra relative to the scalar multiplication determined by*

$$\lambda(\mu \otimes x) = (\lambda\mu) \otimes x, \quad \lambda, \mu \in K, \ x \in A.$$

Proof Take $\lambda \in K$. The map

$$K \times A \to K \otimes A, \quad (\mu, x) \mapsto (\lambda\mu) \otimes x,$$

is bilinear, and hence there is an F-linear map

$$K \otimes_F A \to K \otimes_F A, \ \mu \otimes x \mapsto (\lambda \mu) \otimes x$$

(incidentally, this also follows by using Lemma 4.5 for the (left) multiplication map L_λ on K and the identity map on A). Our scalar multiplication is thus well-defined and $\lambda(r + s) = \lambda r + \lambda s$ holds for $\lambda \in K$, $r, s \in K \otimes_F A$. It is straightforward to check that we also have

$$(\lambda + \mu)r = \lambda r + \mu r, \quad (\lambda \mu)r = \lambda(\mu r), \quad \lambda(rs) = (\lambda r)s = r(\lambda s)$$

for $\lambda, \mu \in K$, $r, s \in K \otimes_F A$. $\qquad \square$

Definition 4.37 Let A be an F-algebra and let K be an extension field of F. The K-algebra $K \otimes_F A$ (defined in Lemma 4.36) is called the **scalar extension** of A to K. It will be denoted by A_K.

The following assertions are immediate:

- $1 \otimes A$ is an F-subalgebra of A_K isomorphic to A.
- A_K is linearly spanned by $1 \otimes A$.
- If $\{e_i \,|\, i \in I\}$ is a basis of A, then $\{1 \otimes e_i \,|\, i \in I\}$ is a basis of A_K (cf. Corollary 4.14).

Assume now that A is finite dimensional. Then we obviously have

$$[A_K : K] = [A : F].$$

If $\{e_1, \ldots, e_d\}$ is a basis of A with multiplication table

$$e_i e_j = \sum_{k=1}^{d} \alpha_{ijk} e_k, \ \alpha_{ijk} \in F,$$

then the elements $\overline{e}_i = 1 \otimes e_i$ form a basis of A_K with the same multiplication table:

$$\overline{e}_i \overline{e}_j = \sum_{k=1}^{d} \alpha_{ijk} \overline{e}_k.$$

Neglecting the insignificant notational differences, we thus see that Definition 4.37 coincides with the definition given in Sect. 2.11. Being independent of basis, the new definition is not only more clear and elegant, but also more suitable for practical use.

We already saw in Sect. 2.11 that scalar extensions can be used to transform finite dimensional central simple algebras into the matrix algebras over fields. Let us consider this phenomenon from the conceptual standpoint.

Definition 4.38 Let A be a finite dimensional central simple F-algebra. An extension field K of F is said to be a **splitting field** for A if $A_K \cong M_n(K)$ for some $n \in \mathbb{N}$.

We can now reword Theorem 2.74 as follows.

Theorem 4.39 *An algebraic closure $\overline{F}$ of F is a splitting field for every finite dimensional central simple F-algebra.*

For division algebras we have another, more intrinsic class of splitting fields:

Theorem 4.40 *Let D be a finite dimensional central division algebra. Every maximal subfield K of D is a splitting field for D.*

Proof If $a \in C_D(K)$, then a and K generate a field. Therefore $a \in K$ since K is maximal. That is, $C_D(K) = K$. Similarly, $C_{D^\circ}(K) = K$. Corollary 4.35 therefore shows that $K \otimes_F D$ and $\text{End}_K(D^\circ)$ are isomorphic as F-algebras. However, from the proof we see that the isomorphism is given by $k \otimes d \mapsto L_k R_d$, and so they are actually isomorphic as K-algebras. Since D° is finite dimensional over K, we have $\text{End}_K(D^\circ) \cong M_n(K)$. $\square$

Example 4.41 A basic example of a maximal subfield of $\mathbb{H}$ is $\mathbb{R} \oplus \mathbb{R}i$; but in view of Theorem 4.39 this is hardly a new example of a splitting field for $\mathbb{H}$. If we take, for instance, the *rational quaternions* $\mathbb{Q} \oplus \mathbb{Q}i \oplus \mathbb{Q}j \oplus \mathbb{Q}k$, then Theorems 4.39 and 4.40 do yield different splitting fields.

4.8 Simplicity of Tensor Products

The basic Wedderburn's structure theorem (Theorem 2.61) can be interpreted as follows: A finite dimensional algebra is simple if and only if it is isomorphic to the tensor product of a matrix algebra $M_n(F)$ and a division algebra D (cf. Example 4.22). This shows that, unlike the direct product, a more sophisticated tensor product may give rise to simple algebras. Let us examine when this can occur.

Theorem 4.42 *Let A be an arbitrary unital algebra, and let B be a finite dimensional central simple algebra. Then every ideal of $A \otimes B$ is of the form $I \otimes B$ where I is an ideal of A. Accordingly, $A \otimes B$ is simple if and only if A is simple.*

Proof Let U be an ideal of $A \otimes B$. It is immediate that $I := \{a \in A \mid a \otimes 1 \in U\}$ is then an ideal of A, and that $I \otimes B \subseteq U$. To prove the converse inclusion, pick $u \in U$. Let $\{f_1, \ldots, f_n\}$ be a basis of B. By Corollary 4.14 we have $u = \sum_{j=1}^n a_j \otimes f_j$ for some $a_j \in A$. Take $\varphi \in \text{End}_F(B)$ such that $\varphi(f_1) = 1$ and $\varphi(f_j) = 0$, $j \geq 2$. In view of Lemma 1.25, we can represent φ as $\sum_{i=1}^r L_{b_i} R_{c_i}$, $b_i, c_i \in B$. Hence we have

$$U \ni \sum_{i=1}^r (1 \otimes b_i)u(1 \otimes c_i) = \sum_{i=1}^r \sum_{j=1}^n a_j \otimes b_i f_j c_i = \sum_{j=1}^n a_j \otimes \varphi(f_j) = a_1 \otimes 1.$$

Thus, $a_1 \in I$. Similarly we see that each $a_i \in I$, and so $u \in I \otimes B$. Thus $U = I \otimes B$.

If A is simple, then $I = 0$ or $I = A$, and hence $U = 0$ or $U = A \otimes B$. Thus $A \otimes B$ is simple. Conversely, assume that $A \otimes B$ is simple, and choose a nonzero ideal I of A. Then $I \otimes B$ is a nonzero ideal of $A \otimes B$, so that $I \otimes B = A \otimes B$. Given $a \in A$, we can therefore find $a_i \in I$ and $b_i \in B$ such that $a \otimes 1 = \sum_i a_i \otimes b_i$. From Lemma 4.11 we see that a is a linear combination of the a_i's. Hence $a \in I$. Thus $I = A$, proving the simplicity of A.   □

The assumption that B is finite dimensional can actually be removed. The proof is basically the same, one just has to apply Corollary 5.24 from the next chapter instead of Lemma 1.25. For our current purposes, however, the finite dimensional case is sufficient.

Corollary 4.43 *Let A be an arbitrary unital algebra. Then every ideal of $M_n(A)$ is of the form $M_n(I)$, where I is an ideal of A. Accordingly, $M_n(A)$ is simple if and only if A is simple.*

Proof Take $M_n(F)$ for B in Theorem 4.42.   □

It should be remarked that Corollary 4.43 can be also proved by entirely elementary means, avoiding the tensor products.

Corollary 4.44 *The tensor product of two finite dimensional central simple algebras is a finite dimensional central simple algebra.*

Proof Use Theorem 4.42 together with Corollary 4.33.   □

The tensor product of two finite dimensional simple algebras may not be simple. If, however, one of them is central, then it is. If both are central, then so is their tensor product. These phenomena are nicely illustrated by the tensor products of real division algebras:

Example 4.45 We have

$$\mathbb{C} \otimes_{\mathbb{R}} \mathbb{C} \cong \mathbb{C} \times \mathbb{C}, \quad \mathbb{C} \otimes_{\mathbb{R}} \mathbb{H} \cong M_2(\mathbb{C}), \quad \mathbb{H} \otimes_{\mathbb{R}} \mathbb{H} \cong M_4(\mathbb{R}).$$

The last statement was noticed in Example 4.29. We know already from Sect. 2.11 that the scalar extension of $\mathbb{H}$ to $\mathbb{C}$ is $M_2(\mathbb{C})$. Therefore $\mathbb{C} \otimes_{\mathbb{R}} \mathbb{H}$ is isomorphic to $M_2(\mathbb{C})$ even when we consider them as $\mathbb{C}$-algebras (and hence also when we consider them as $\mathbb{R}$-algebras). We may also regard $\mathbb{C} \otimes_{\mathbb{R}} \mathbb{C}$ as a $\mathbb{C}$-algebra (in light of Lemma 4.36). As such it is isomorphic to $\mathbb{C} \times \mathbb{C}$ since they are both unital 2-dimensional $\mathbb{C}$-algebras containing an element which is not a scalar multiple of unity and whose square is -1 (i.e., $1 \otimes i$ and $(i, -i)$, respectively). Therefore $\mathbb{C} \otimes_{\mathbb{R}} \mathbb{C}$ and $\mathbb{C} \times \mathbb{C}$ are isomorphic also as $\mathbb{R}$-algebras.

4.9 The Skolem-Noether Theorem

Most parts of this and the preceding chapter were devoted to building machinery. In these final sections we shall finally be rewarded. We are going to prove two appealing theorems on finite dimensional central simple algebras by using several abstract results on tensor products and modules.

In Sect. 1.6 we have proved that all automorphisms of finite dimensional central simple algebras are inner. As we pointed out then, this result is an important special case of the so-called Skolem-Noether Theorem. We are now in a position to establish the general version of this theorem.

Theorem 4.46 (Skolem-Noether) *Let A be a finite dimensional central simple algebra. If S is a simple subalgebra of A that contains the unity 1 of A, then every homomorphism from S into A that maps 1 into 1 can be extended to an inner automorphism of A.*

Proof As we know, A is isomorphic to a matrix algebra over a central division algebra (Corollary 2.62). However, for the purposes of this proof it is more convenient to represent it as an endomorphism algebra. More precisely, according to Theorem 3.31 we have $A \cong \mathrm{End}_\Delta(V)$, where Δ is a central division algebra and V is a finite dimensional vector space over Δ. In fact, there is no loss of generality in assuming that $A = \mathrm{End}_\Delta(V)$ (note that the conclusion of the theorem holds simultaneously for isomorphic algebras).

Let $T = S \otimes_F \Delta$, where F is the base field. As S is simple and Δ is central simple, T is a simple algebra by Theorem 4.42. Let us point out that V is also a vector space over F and elements in A are F-linear maps (cf. Remark 3.34). Given $v \in V$, we see that the map $S \times \Delta \to V$, $(\sigma, \delta) \mapsto \sigma(\delta v)$ is bilinear, and hence there is an F-linear map $T \to V$, $\sigma \otimes \delta \mapsto \sigma(\delta v)$. Since v is arbitrary, there exists a bilinear map $T \times V \to V$, $(t, v) \mapsto tv$, such that

$$(\sigma \otimes \delta)v = \sigma(\delta v) \quad \text{for all } \sigma \in S, \delta \in \Delta, v \in V. \tag{4.3}$$

In this way V, considered as a space over F, becomes a unital T-module. Indeed, since elements in S are Δ-linear maps, we have

$$((\sigma \otimes \delta)(\sigma' \otimes \delta'))v = \sigma\sigma'((\delta\delta')v) = \sigma\Big(\sigma'(\delta(\delta'v))\Big)$$
$$= \sigma\Big(\delta(\sigma'(\delta'v))\Big) = (\sigma \otimes \delta)((\sigma' \otimes \delta')v).$$

To avoid confusion, we denote V, when viewed as a T-module with respect to (4.3), by V_1.

Let $\varphi : S \to A$ be a homomorphism such that $\varphi(1) = 1$. By making only obvious changes in the above discussion we see that V becomes a unital T-module with respect to the action

$$(\sigma \otimes \delta) \cdot v = \varphi(\sigma)(\delta v) \quad \text{for all} \quad \sigma \in S, \delta \in \Delta, v \in V. \tag{4.4}$$

Let us denote V, when viewed as a T-module with respect to (4.4), by V_2.

We have arrived at the crucial point of the proof: since V_1 and V_2 have the same dimension over F (indeed, as vector spaces over F they are simply identical) and T is a simple algebra, Corollary 3.60 implies that they are isomorphic as T-modules. Let $\alpha : V_1 \to V_2$ be an isomorphism. Given $\delta \in \Delta$ and $v \in V$, we have

$$\alpha(\delta v) = \alpha((1 \otimes \delta)v) = 1 \otimes \delta \cdot \alpha(v) = \varphi(1)(\delta\alpha(v)) = \delta\alpha(v).$$

That is, α is a Δ-linear map. As vector spaces over F, both V_1 and V_2 are equal to V. Therefore $\alpha \in A$. Moreover, since α is bijective, it is invertible in A.

Let $\sigma \in S$. We can multiply α and σ as elements in A. For $v \in V$ we have

$$(\alpha\sigma)(v) = \alpha(\sigma(v)) = \alpha((\sigma \otimes 1)v) = \sigma \otimes 1 \cdot \alpha(v) = \varphi(\sigma)(\alpha(v)) = (\varphi(\sigma)\alpha)(v).$$

Since v is arbitrary, this means that $\alpha\sigma = \varphi(\sigma)\alpha$ for every $\sigma \in S$. That is, $\varphi(\sigma) = \alpha\sigma\alpha^{-1}$, so φ can be extended to the inner automorphism $a \mapsto \alpha a \alpha^{-1}$ of A. $\square$

The Skolem-Noether Theorem turns out to be very useful. In particular, as one can find in several textbooks (e.g., in [FD93], [Her68], and [Hun74]). Frobenius' theorem on real division algebras and Wedderburn's theorem on finite division rings can be derived from it. Of course, these proofs are not as elementary as those given in Chap. 1. But they are very elegant and therefore worth seeing.

4.10 The Double Centralizer Theorem

Let S be a subalgebra of an algebra A. The centralizer of its centralizer, $C_A(C_A(S))$, is called the **double centralizer** of S. This is obviously a subalgebra of A which contains S. Quite possibly S is its proper subset. Say, if S does not contain $Z(A)$, this is certainly the case. Another example: if $A = M_n(F)$ and $S = T_n(F), n \geq 2$, then $C_A(S) = F = Z(A)$, and hence $C_A(C_A(S)) = A \supsetneq S$. On the other hand, if $S = A$ or $S = Z(A)$, then $C_A(C_A(S)) = S$. A more illuminating situation where the equality holds is described in the following example.

Example 4.47 Considering central algebras S and T as subalgebras of $A := S \otimes T$, we have $C_A(S) = T$ and $C_A(T) = S$ by Corollary 4.33, and hence $C_A(C_A(S)) = S$ and $C_A(C_A(T)) = T$. If S and T are finite dimensional, then we also have $[A : F] = [S : F][C_A(S) : F] = [T : F][C_A(T) : F]$.

This example illustrates the next theorem in which, however, we will not assume that S is central.

Theorem 4.48 (Double Centralizer Theorem) *Let A be a finite dimensional central simple F-algebra. If S is a simple subalgebra of A that contains the unity of A, then $C_A(S)$ is a simple algebra,*

$$[A : F] = [S : F][C_A(S) : F], \tag{4.5}$$

and $C_A(C_A(S)) = S$.

Proof The last claim follows from the first two. Namely, knowing that $C_A(S)$ is simple, we can replace S by $C_A(S)$ in (4.5). Thus,

$$[A : F] = [C_A(S) : F][C_A(C_A(S)) : F].$$

Comparing this formula with (4.5) we obtain $[C_A(C_A(S)) : F] = [S : F]$. Since $S \subseteq C_A(C_A(S))$, the desired conclusion $C_A(C_A(S)) = S$ follows.

We divide the proof of the first two claims into two steps. In the first one we will consider the case where S is a division algebra, and in the second one we will consider the general case.

(I) Assume that $S = D$ is a division algebra. By Corollary 4.35 we have $C_A(D) \otimes A^\circ \cong \mathrm{End}_D(A)$. Theorem 3.31 shows that $\mathrm{End}_D(A) \cong M_s(D^\circ)$, where $s = \dim_D A$. Therefore $\mathrm{End}_D(A)$ is a simple algebra. Using the easier "only if" part of Theorem 4.42 it follows that $C_A(D)$ is simple. It remains to prove the second claim. On the one hand, we have

$$[\mathrm{End}_D(A) : F] = s^2[D : F],$$

and on the other hand,

$$[\mathrm{End}_D(A) : F] = [C_A(D) : F][A : F].$$

We claim that

$$s = \frac{[A : F]}{[D : F]}.$$

If D was a field, this would follow from Remark 1.33. Fortunately, the proof remains unchanged if D is noncommutative. Combining the above three formulas we obtain (4.5) (for $S = D$).

(II) Assume now that S is an arbitrary simple subalgebra of A. As we know from Wedderburn's theorem (Theorem 2.61), there exist $r \in \mathbb{N}$ and a division algebra D such that $S \cong M_r(D)$. This implies that S contains a set of $r \times r$ matrix units e_{ij}. These are also the matrix units of A for A has the same unity as S. Lemma 2.52 therefore implies that $A \cong M_r(B)$, where $B = e_{11}Ae_{11} \supseteq e_{11}Se_{11} = D$. Since B is a central algebra (Lemma 1.15), we see from Corollary 4.34 that

$$C_A(S) \cong C_B(D).$$

As B is also a simple algebra (Corollary 4.43), we can apply (**I**) with B playing the role of A. Thus, $C_B(D)$ is a simple algebra and $[B : F] = [D : F][C_B(D) : F]$. The former establishes the simplicity of $C_A(S)$, and the latter implies

$$[A : F] = r^2[B : F] = r^2[D : F][C_B(D) : F].$$

Since $r^2[D : F] = [S : F]$ and $[C_B(D) : F] = [C_A(S) : F]$, this proves (4.5). $\square$

Adding the assumption that S is central, we arrive at the situation considered in Example 4.47.

Corollary 4.49 *Let A be a finite dimensional central simple F-algebra. If S is a central simple subalgebra of A that contains the unity of A, then $A \cong S \otimes C_A(S)$.*

Proof Consider a linear map $\varphi : S \otimes C_A(S) \to A$, $\varphi(s \otimes t) = st$. Since elements from S commute with elements from $C_A(S)$, φ is multiplicative. The algebra $S \otimes C_A(S)$ is simple (Theorem 4.42), so that $\ker \varphi = 0$. From $[A : F] = [S : F][C_A(S) : F]$ we see that $S \otimes C_A(S)$ and A are of the same dimension. Hence φ is an isomorphism. $\square$

Corollary 4.49 deals with the case where the center of S is minimal. Let us take a look at the opposite extreme. Assume that $S = K$ is a maximal subfield of a finite dimensional central division algebra D. As observed in the proof of Theorem 4.40, $C_D(K) = K$ in this case. Corollary 4.49 does not hold for $S = K$; after all, $K \otimes K$ is a commutative algebra. The only nontrivial information that we extract from the Double Centralizer Theorem is $[D : F] = [K : F]^2$, which is nothing but the content of Theorem 1.36.

4.11 The Brauer Group

The basic idea behind the concept of the Brauer group is to use the tensor products to classify finite dimensional central division algebras over a given field. The tensor product of two such algebras is not always a division algebra. For instance, $\mathbb{H} \otimes_{\mathbb{R}} \mathbb{H} \cong M_4(\mathbb{R})$ (Example 4.29). However, it always is a central simple algebra (Corollary 4.44), and such an algebra conceals within itself a central division algebra. That is, it is isomorphic, for some $n \in \mathbb{N}$, to $M_n(D)$, where D is a central division algebra (Corollary 2.62). Moreover, D is essentially unique (Corollary 3.58). Thus, ignoring the matrix part of the resulting central simple algebra, we can say that the tensor product of two finite dimensional central division algebras gives rise to an algebra from the same class. Say, the product of $\mathbb{H}$ with itself brings out $\mathbb{R}$. (Maybe not a very exciting result, but what to expect in view of Frobenius' theorem?)

Let us now make everything precise. Fix a field F. Let A and A' be finite dimensional central simple F-algebras. Then, as we know, there exist $n, n' \in \mathbb{N}$ and central division algebras D, D' such that $A \cong M_n(D)$ and $A' \cong M_{n'}(D')$. We shall say that A and A' are **equivalent**, and write $A \sim A'$, if $D \cong D'$. This definition is unambiguous since D and D' are uniquely determined up to isomorphism. Note that $\sim$ is indeed an equivalence relation in the class of all finite dimensional central simple F-algebras. The equivalence class of A will be denoted by $[A]$. Thus, if D is a finite dimensional central division algebra, then $[D] = [M_2(D)] = [M_3(D)] = \ldots$. As indicated in the previous paragraph, among all representatives of this class we are really interested only in D (and division algebras isomorphic to D). The problem is that D may be hidden, so we are forced to treat all representatives equally.

Proposition 4.50 *The set of all equivalence classes of finite dimensional central simple F-algebras, endowed with multiplication*

$$[A] \cdot [B] = [A \otimes B],$$

is an abelian group.

Proof We first have to show that this multiplication is well-defined. That is, we have to prove that for all finite dimensional central simple algebras A, A', B, B',

$$A \sim A' \text{ and } B \sim B' \implies A \otimes B \sim A' \otimes B'.$$

By definition, $A \sim A'$ means $A \cong M_n(D)$ and $A' \cong M_{n'}(D')$ with $D \cong D'$. But then we actually have $A' \cong M_{n'}(D)$. Similarly, $B \sim B'$ implies that $B \cong M_m(C)$ and $B' \cong M_{m'}(C)$ for some $m, m' \in \mathbb{N}$ and a central division algebra C. Since $D \otimes C$ is again a finite dimensional central simple algebra by Corollary 4.44, we have $D \otimes C \cong M_k(E)$ with $k \in \mathbb{N}$ and E a central division algebra. Now, using the associativity and commutativity of tensor products, together with basic facts about matrix algebras represented as tensor products (explained in Example 4.22), we obtain

$$\begin{aligned}
A \otimes B &\cong \big(M_n(F) \otimes D\big) \otimes \big(M_m(F) \otimes C\big) \\
&\cong \big(M_n(F) \otimes M_m(F)\big) \otimes \big(D \otimes C\big) \\
&\cong M_{nm}(F) \otimes \big(M_k(F) \otimes E\big) \\
&\cong M_{nmk}(F) \otimes E \\
&\cong M_{nmk}(E).
\end{aligned}$$

Similarly we see that $A' \otimes B' \cong M_{n'm'k}(E)$. Accordingly, $A \otimes B \sim A' \otimes B'$.

The multiplication is associative and commutative since the tensor product is associative and commutative. Obviously, $[F]$ is the identity element. Finally, from $A \otimes A^\circ \cong M_n(F)$ (Corollary 4.28) we infer that $[A]^{-1} = [A^\circ]$. $\square$

Some of the readers may have noticed that there is a missing point in this proof. One would first have to verify that the equivalence classes of finite dimensional central simple F-algebras indeed form a set. This is not a difficult task, but we shall not go into this matter.

Definition 4.51 The group from Proposition 4.50 is called the **Brauer group** of the field F. It will be denoted by $\mathbf{Br}(F)$.

The name was given after R. Brauer who made fundamental contributions to this field of research.

Computing $\mathbf{Br}(F)$ for a given field F is not an easy task. But at least we know the following:

- If F is algebraically closed, then $\mathbf{Br}(F) = \{1\}$.
- If F is finite, then $\mathbf{Br}(F) = \{1\}$.
- $\mathbf{Br}(\mathbb{R}) \cong \mathbb{Z}_2$.

These are restatements of Proposition 1.6, Theorem 1.38, and Corollary 1.19 with Example 4.29.

Brauer groups turn out to be important in different mathematical areas. Anyhow, we close our discussion on this topic at this point, and refer to [FD93] and [Pie82] for possible further reading. One might protest that we have just warmed up, and besides the definition and fancy interpretations of some fundamental results from Chap. 1 we have told nothing. But perhaps the reader will agree that the definition of the Brauer group is captivating in its own right. This short section is only an invitation into a rich area.

Exercises

4.1. Let U and V be vector spaces of dimension at least 2. Prove that not every element in $U \otimes V$ is a simple tensor.

4.2. Let U be a real vector space. Show that the identity $\sum_{i=1}^{n} u_i \otimes u_i = 0$ in $U \otimes U$ implies that each $u_i = 0$.

4.3. Let U be a vector space over a field F with char$(F) \neq 2$. Denote by S the subspace of $U \otimes U$ generated by all elements of the form $u \otimes u$, $u \in U$, and by T the subspace generated by all elements of the form $u \otimes v - v \otimes u$, $u, v \in U$. Show that $U \otimes U = S \oplus T$. What is the dimension of S and T if $\dim_F U = n < \infty$?

Hint: Make use of the map $u \otimes v \mapsto v \otimes u$.

4.4. Show that $(U_1 \oplus U_2) \otimes V \cong (U_1 \otimes V) \oplus (U_2 \otimes V)$ holds for vector spaces. Extend this to algebras.

4.5. Use the tensor product to construct a 9-dimensional algebra whose radical is 5-dimensional.

4.6. Let $S = (s_{ij}) \in M_m(F)$ and $T = (t_{ij}) \in M_n(F)$. The $mn \times mn$ matrix

$$S \otimes T := \begin{bmatrix} s_{11}T & s_{12}T & \cdots & s_{1m}T \\ s_{21}T & s_{22}T & \cdots & s_{2m}T \\ \vdots & \vdots & \ddots & \vdots \\ s_{m1}T & s_{m2}T & \cdots & s_{mm}T \end{bmatrix} = \begin{bmatrix} s_{11}t_{11} & s_{11}t_{12} & \cdots & s_{1m}t_{1n} \\ s_{11}t_{21} & s_{11}t_{22} & \cdots & s_{1m}t_{2n} \\ \vdots & \vdots & \ddots & \vdots \\ s_{m1}t_{n1} & s_{m1}t_{n2} & \cdots & s_{mm}t_{nn} \end{bmatrix}$$

is called the **Kronecker** or **tensor product** of matrices S and T. This definition is clearly in accordance with the convention from Example 4.22. Note that it is also compatible with the tensor product of linear maps in the following sense: If S is the matrix corresponding to a linear map f with respect to a basis $\{e_1, \ldots, e_m\}$, and T is the matrix corresponding to a linear map g with respect to a basis $\{f_1, \ldots, f_n\}$, then $S \otimes T$ is the matrix corresponding to $f \otimes g$ with respect to the basis $\{e_1 \otimes f_1, \ldots, e_1 \otimes f_n, e_2 \otimes f_1, \ldots, e_m \otimes f_n\}$. Show that $\mathrm{tr}(S \otimes T) = \mathrm{tr}(S)\mathrm{tr}(T)$ and $\det(S \otimes T) = \det(S)^n \det(T)^m$.

Hint: $S \otimes T = (S \otimes I_n)(I_m \otimes T)$ where I_k stands for the $k \times k$ identity matrix.

4.7. Show that algebras A and B must be unital in case $A \otimes B$ is unital.

4.8. Show that $p \in \mathbb{N}$ is prime if and only if there do not exist algebras $A \neq F$ and $B \neq F$ such that $A \otimes B \cong M_p(F)$.

4.9. Let A, A', B, B' be finite dimensional simple algebras. Suppose there exist $m, n \in \mathbb{N}$ such that $M_m(A) \cong M_n(B)$ and $M_n(A') \cong M_m(B')$. Show that $A \otimes A' \cong B \otimes B'$.

4.10. Let A be a central simple algebra over a field F with $\mathrm{char}(F) \neq 2, 3$. Show that $a \in A$ satisfies $[a, [a, [a, x]]] = 0$ for all $x \in A$ if and only if there exists $\lambda \in F$ such that $(a - \lambda)^2 = 0$. Analyze, in a similar fashion, the condition $[a, [a, [a, x]]] = [a, x]$.

Hint: Interpret the conditions $[a, [a, [a, x]]] = 0$ and $[a, [a, [a, x]]] = [a, x]$ in the multiplication algebra $M(A)$.

4.11. Let A be a 4-dimensional central simple F-algebra. Show that $C_A(\{a\}) = F + Fa$ for every $a \in A \setminus F$.

Hint: Apply scalar extension to reduce the problem to the case where $A = M_2(F)$ and F is algebraically closed (so that the Jordan normal form of a is applicable).

4.12. Let A be a finite dimensional central simple algebra. Show that the linear span of all commutators in A has codimension 1 in A.

4.13. Let A be an F-algebra and let K be an extension field of F. True or False:

(a) If A is prime, then A_K is prime.
(b) If A_K is prime, then A is prime.

4.14. Let A be an $\mathbb{R}$-algebra. Show that the set $A \times A$ equipped with operations

$$(x, y) + (z, w) = (x + z, y + w),$$
$$(\alpha + i\beta)(x, y) = (\alpha x - \beta y, \beta x + \alpha y),$$
$$(x, y)(z, w) = (xz - yw, xw + yz),$$

where $x, y, z, w \in A$ and $\alpha, \beta \in \mathbb{R}$, is a $\mathbb{C}$-algebra isomorphic to $A_{\mathbb{C}}$.

Remark: This construction provides a shortcut to the notion of the scalar extension for mathematicians working in areas in which $\mathbb{R}$ and $\mathbb{C}$ are the only fields that matter.

4.15. Let D_1 and D_2 be finite dimensional central division F-algebras. As we know, $D_1 \otimes D_2$ is isomorphic to $M_m(D)$ for some $m \in \mathbb{N}$ and a central division algebra D. Examine $D_i^o \otimes M_m(D)$ to show that m divides $[D_i : F], i = 1, 2$. Hence conclude that $D_1 \otimes D_2$ is a division algebra in case $[D_1 : F]$ and $[D_2 : F]$ are relatively prime.

Remark: The latter fact is usually used in the proof of the following important and beautiful result: If D is a finite dimensional central division F-algebra, and $[D : F] = p_1^{k_1} \ldots p_r^{k_r}$ where p_i are different primes, then there exist central division algebras $D_1, \ldots, D_r$ such that $D \cong D_1 \otimes \cdots \otimes D_r$ and $[D_i : F] = p_i^{k_i}$.

4.16. Let $\mathrm{char}(F) = p > 0$. Suppose a unital F-algebra A contains an element a such that $a \notin F$ and $a^p \in F$. Show that then $A \otimes A$ contains a nonzero nilpotent element.

Remark: Using this one can show that even the tensor product of fields is not always a semiprime algebra. On the other hand, if $\mathrm{char}(F) = 0$ and K, L are field extensions of F such that at least one of them is finite, then $K \otimes_F L$ is a direct product of fields (and is hence semiprime). The proof uses only standard field extension tools, and readers can try to produce it themselves.

4.17. Let A and B be simple unital F-algebras. Show that the algebra $A \otimes B$ is semiprime if (and only if) its center is semiprime.

Hint: Take a nilpotent ideal I of $A \otimes B$, and consider an element $\sum_{i=1}^{n} a_i \otimes b_i \neq 0$ in I such that n is minimal.

4.18. Show that besides $0, \mathbb{R}$, and $\mathbb{H}$, the only subalgebras of $\mathbb{H}$ are those of the form $h(\mathbb{R} \oplus \mathbb{R}i)h^{-1}, h \in \mathbb{H} \setminus \{0\}$.

4.19. Let A be a finite dimensional central simple algebra over a field F with the property that $\lambda^2 \neq -1$ for every $\lambda \in F$. Show that if $u, v \in A$ satisfy $u^2 = v^2 = -1$, then there exists an invertible $a \in A$ such that $u = ava^{-1}$.

Hint: $F + Fu$ and $F + Fv$ are isomorphic subfields of A.

4.20. A linear map δ from an algebra A into itself is called a **generalized derivation** if $\delta(xyz) = \delta(xy)z - x\delta(y)z + x\delta(yz)$ for all $x, y, z \in A$. Show that every generalized derivation of a finite dimensional central simple algebra A is the sum of a left multiplication map and a right multiplication map.

Hint: Observe that

$$\begin{bmatrix} x & 0 \\ 0 & x \end{bmatrix} \mapsto \begin{bmatrix} x & \delta(x) - \delta(1)x \\ 0 & x \end{bmatrix}$$

defines an algebra homomorphism from an isomorphic copy of A in $M_2(A)$ into $M_2(A)$.

4.21. Let S be a subalgebra of an algebra A. Show that:

(a) If S is commutative, then so is $C_A(C_A(S))$, and $C_A(C_A(S)) \subseteq C_A(S)$.
(b) If S is equal to the centralizer of some subset of A, then $S = C_A(C_A(S))$.

4.22. Let A and S be as in the Double Centralizer Theorem. Note that the center Z of S is a field, and that S, $C_A(S)$, and $C_A(Z)$ are algebras over Z. Show that $S \otimes_Z C_A(S) \cong C_A(Z)$.

4.23. Let A be a finite dimensional central simple $\mathbb{R}$-algebra. Show that A contains a subfield K such that $K = C_A(K)$ if and only if A is isomorphic to $\mathbb{R}$, $M_2(\mathbb{R})$, or $\mathbb{H}$.

4.24. Let A and A' be finite dimensional central simple F-algebras. Show that:

(a) $A \sim A'$ if and only if there exist $m, m' \in \mathbb{N}$ such that $M_m(A) \cong M_{m'}(A')$.
(b) $A \cong A'$ if and only if $A \sim A'$ and $[A : F] = [A' : F]$.

Chapter 5
Structure of Rings

Although most of the concepts introduced in the first four chapters involve general rings and algebras, the majority of the main results concern algebras of finite dimension. In this chapter we will make a clean break from the finite dimensional context.

The goal of the chapter is to extend Wedderburn's structure theory of finite dimensional algebras, discussed in Chap. 2, to general rings. As one would expect, this cannot be done in a straightforward manner and the main results are considerably more complex.

The topics will be considered in a different order than in Chap. 2. In the first part we will treat primitive rings, which play a similar role in general rings as simple algebras do in the context of finite dimensional algebras. The second part will be primarily devoted to the Jacobson radical, a generalization of the radical of a finite dimensional algebra.

Most of the theory that we are about to expose was developed around 1945 by N. Jacobson, a student of Wedderburn.

5.1 Primitive Rings

Wedderburn's basic structure theorem says that a finite dimensional prime algebra can be represented as an algebra of all $n \times n$ matrices over a division algebra D, or, equivalently, as an algebra of all linear operators on an n-dimensional vector space over a division algebra $\Delta = D^o$ (cf. Remark 3.34). We are going to introduce a class of prime rings for which a more general structure theorem holds.

The fact that minimal left ideals give rise to division rings (Lemma 2.58) was one of the main ingredients in our proof of Wedderburn's theorem. Relying on minimal left ideals in more general rings would have a limited reach since their existence is questionable. But there is a perfect substitute for them, namely, the simple modules—they exist under rather mild conditions (cf. Lemma 3.46), and they also

© Springer International Publishing Switzerland 2014
M. Brešar, *Introduction to Noncommutative Algebra*, Universitext,
DOI 10.1007/978-3-319-08693-4_5

yield division rings by Schur's Lemma. At this point we advise the reader to glance through Sects. 3.5–3.7 in order to better understand the discussion that follows.

Let R be a ring and let V be a simple (left) R-module. Schur's Lemma (Lemma 3.50) tells us that

$$\Delta := \mathrm{End}_R(V)$$

is a division ring. Let $\delta \in \Delta$ and $v \in V$. We simplify the notation by setting

$$\delta v := \delta(v) \in V. \tag{5.1}$$

Obviously, we have

$$(\delta_1 + \delta_2)v = \delta_1 v + \delta_2 v, \quad \delta(v_1 + v_2) = \delta v_1 + \delta v_2,$$
$$(\delta_1 \delta_2)v = \delta_1(\delta_2 v), \quad 1v = v$$

for all $\delta_1, \delta_2, \delta \in \Delta$ and $v_1, v_2, v \in V$. This means that V is a vector space over Δ; more precisely, the additive group of V becomes a vector space over Δ relative to the operation $\Delta \times V \to V$ defined in (5.1) (that is why we have denoted our simple module by V instead of, say, M). From now on we shall therefore consider V simultaneously as a module over R and a vector space over Δ. Since $\delta \in \Delta$ is an endomorphism of the R-module V, the following fundamental formula

$$\delta(rv) = r(\delta v) \quad \text{for all } \delta \in \Delta, r \in R, v \in V, \tag{5.2}$$

connects both structures. Now, δ and r appear symmetrically in (5.2). Accordingly, why not interpret (5.2) as that r is an endomorphism of the vector space V over Δ? We will formalize this idea in the next section. However, an extra condition is needed for its full realization. If r lies in the annihilator of V, then r gives rise to the zero endomorphism of V over Δ. We will therefore require that V is faithful.

Definition 5.1 A ring R is said to be **primitive** if it has a faithful simple module V.

Recall from Definition 3.11 and Lemma 3.36 that this means that $V \neq 0$ and the following two conditions are fulfilled:

- $rV \neq 0$ for every $0 \neq r \in R$.
- $Rv = V$ for every $0 \neq v \in V$.

Since our modules are, by convention, left modules, a more adequate name for such a ring is a **left primitive ring**. Analogously one defines a **right primitive ring**, and it turns out that these two classes of rings do not coincide. It is good to keep this in mind, but we will omit writing "left" before "primitive ring".

Remark 5.2 Let R be a primitive ring. If a ring S is isomorphic to R, then S is primitive too. Indeed, if V is a faithful simple R-module, then the additive group of V becomes a faithful simple S-module by defining $sv := \varphi(s)v$, where φ is an isomorphism from S onto R.

The discussion before Definition 5.1 indicates a simple example, in fact the prototype, of a primitive ring:

Example 5.3 Let V be a nonzero vector space over a division ring Δ. Then the ring $R := \mathrm{End}_\Delta(V)$ is primitive. Indeed, V is its faithful simple module (via $fv := f(v)$). The condition that $Rv = V$ if $v \neq 0$ follows from Remark 3.29 and the fact that v is contained in a basis (cf. Theorem 3.28). Let us stress that here V is of arbitrary dimension, finite or infinite.

More concrete examples of primitive rings will be met later. Let us now compare the class of primitive rings to other classes of rings introduced earlier.

Lemma 5.4 *A primitive ring R is prime.*

Proof Let $a, b \in R$ be nonzero elements. Take a faithful simple R-module V. By faithfulness, there are $u, v \in V$ such that $au \neq 0$ and $bv \neq 0$, and by simplicity there is $r \in R$ such that $r(bv) = u$. Accordingly, $(arb)v \neq 0$, and so $aRb \neq 0$. $\square$

Lemma 5.5 *A prime ring R with a minimal left ideal is primitive.*

Proof Let L be a minimal left ideal of R. It is clear that L is simple as an R-module. The primeness of R implies that it is also faithful. Indeed, $rL = 0$, where $r \in R$, implies $rRL \subseteq rL = 0$, and hence $r = 0$. $\square$

Lemma 5.6 *A simple unital ring R is primitive.*

Proof Since R is unital, it has simple modules by Corollary 3.49. The annihilator of every module is an ideal, and so simple modules over simple rings are automatically faithful. $\square$

A non-unital simple ring is not always primitive. Such rings may have surprising properties. For example, in 2002 A. Smoktunowicz constructed a simple nil ring. Primitive rings, however, cannot be nil. In fact, if R is a nil ring and M is an R-module, then we clearly have $rm \neq m$ for every $r \in R$ and $0 \neq m \in M$. Thus, in view of Lemma 3.36, nil rings do not even have simple modules.

Lemma 5.7 *A commutative primitive ring R is a field.*

Proof Let V be a faithful simple R-module, and pick $0 \neq v \in V$. We claim that for every $r \in R$, $rv = 0$ forces $r = 0$. Indeed, $Rv = V$ by simplicity of V, hence $rV = rRv = Rrv = 0$, and therefore $r = 0$ by the faithfulness of V. Now take $0 \neq a \in R$. Then $av \neq 0$, and therefore, by the simplicity of V, there exists $b \in R$ such that $b(av) = v$. Consequently, $(xba - x)v = 0$ for every $x \in R$. Therefore $bax = xba = x$, and so $ba = 1$ (in particular, R is unital). As R is commutative, this means that $b = a^{-1}$. $\square$

Example 5.8 The simplest example of a prime ring that is not primitive is $\mathbb{Z}$.

Example 5.9 The Weyl algebra $\mathscr{A}_1$ is a simple unital ring (see Example 1.13), and hence a primitive ring.

Example 5.10 If V is an infinite dimensional vector space over a division ring Δ, then the ring $\operatorname{End}_\Delta(V)$ is primitive, but not simple. An example of its proper nonzero ideal is the set of all **finite rank operators**. These are operators whose **rank**, i.e., the dimension of their image, is finite. It is easy to check that finite rank operators indeed form an ideal. It is proper since it does not contain the identity operator, and it certainly is nonzero; in fact, from Remark 3.29 we deduce that for every subspace W of V there is a linear operator having W as its image.

The ring $\operatorname{End}_\Delta(V)$ is also the simplest example of a primitive ring that has minimal left ideals. An example of a primitive ring without minimal left ideals is the Weyl algebra $\mathscr{A}_1$. All this is easy to see, but we postpone the explanation until Sect. 5.4.

Let us close this section by mentioning that every F-algebra can be embedded into a primitive ring, namely, into $\operatorname{End}_F(V)$ for some vector space V over F (Proposition 2.38). This is not always true for rings. For example, take a ring that contains nonzero elements u and v satisfying $2u = 3v = 0$ (e.g., $\mathbb{Z}_2 \times \mathbb{Z}_3$). In light of Remark 2.20 we see that this ring cannot be embedded into a prime ring, and hence neither into a primitive ring by Lemma 5.4.

5.2 The Jacobson Density Theorem

The following notion is often one's first thought when encountering a primitive ring.

Definition 5.11 Let V be a vector space over a division ring Δ, and let R be a subring of the endomorphism ring $\operatorname{End}_\Delta(V)$. We say that R is a **dense ring of linear operators** of V if for every $n \in \mathbb{N}$, every linearly independent subset $\{u_1, u_2, \ldots, u_n\}$ of V, and every (not necessarily linearly independent) subset $\{v_1, v_2, \ldots, v_n\}$ of V, there exists $f \in R$ such that

$$f(u_1) = v_1, \, f(u_2) = v_2, \ldots, f(u_n) = v_n.$$

The term "dense" comes from a topological characterization. We will not need it, so let us not go into detail about this. Still, having in mind the meaning of density in topology, we may have got the message; a dense ring of linear operators of V can be considered as a rather large subring of $\operatorname{End}_\Delta(V)$, or better, large enough for certain purposes.

Lemma 5.12 *A ring R is a dense ring of linear operators of the vector space V over Δ if and only if for every $g \in \operatorname{End}_\Delta(V)$ and every finite dimensional subspace U of V there exists $f \in R$ such that $f|_U = g|_U$.*

Proof Take $g \in \mathrm{End}_\Delta(V)$ and a finite dimensional subspace U of V. Pick a basis $\{u_1, u_2, \ldots, u_n\}$ of U. If R is a dense ring, then there exists $f \in R$ such that $f(u_i) = g(u_i), i = 1, \ldots, n$. Consequently, $f|_U = g|_U$.

Proving the converse is just as easy, only standard facts about endomorphisms and bases have to be used (cf. Remark 3.29). $\qquad\square$

Corollary 5.13 *If V is a finite dimensional vector space over Δ, then $\mathrm{End}_\Delta(V)$ is the only dense ring of linear operators of V.*

If V is infinite dimensional, then we have a variety of dense rings of operators on V. The next lemma gives a partial evidence for this.

Lemma 5.14 *If R is a dense ring of linear operators of a vector space V, then so are its nonzero ideals.*

Proof Let I be a nonzero ideal of R. Choose a nonzero $h \in I$, and let $u \in V$ be such that $v := h(u) \neq 0$. Now take linearly independent vectors $u_1, \ldots, u_n \in V$ and arbitrary vectors $v_1, \ldots, v_n \in V$. Since R is dense, there exist $g_1, \ldots, g_n \in R$ such that $g_i(u_i) = u$ and $g_i(u_j) = 0$ if $j \neq i$. Further, there are $f_1, \ldots, f_n \in R$ such that $f_i(v) = v_i$. Consequently,

$$k := f_1 h g_1 + \cdots + f_n h g_n \in I$$

satisfies $k(u_i) = v_i, i = 1, \ldots, n$. $\qquad\square$

Example 5.15 The ring of all finite rank operators on an infinite dimensional vector space V over Δ is thus a proper dense subring of $\mathrm{End}_\Delta(V)$ (cf. Example 5.10).

The next theorem is one of the cornerstones of noncommutative algebra.

Theorem 5.16 (Jacobson Density Theorem) *A ring R is primitive if and only if it is isomorphic to a dense ring of linear operators of a vector space over a division ring.*

Proof Let us first resolve the easy part. A dense ring of linear operators of a vector space V is primitive, since V is its faithful simple module (via $fv := f(v)$). In fact, one only needs the $n = 1$ case of Definition 5.11 to establish this. In view of Remark 5.2, the "if" part is thus proved.

We proceed to the nontrivial part. Let R be a primitive ring and let V be its faithful simple module. Then $\Delta = \mathrm{End}_R(V)$ is a division ring by Schur's Lemma. The additive group of V becomes a vector space over Δ with respect to the operation defined by (5.1) (i.e., $\delta v := \delta(v)$), and the formula (5.2) (i.e., $\delta(rv) = r(\delta v)$) holds. All this is easy to see and was explained in greater detail in the previous section. Now, for each $r \in R$ we define $\bar{r} : V \to V$ according to

$$\bar{r}(v) := rv. \tag{5.3}$$

It is clear that $\bar{r}$ is an additive map. Moreover, by (5.2) we have $\bar{r} \in \mathrm{End}_\Delta(V)$. Note that

$$R \to \mathrm{End}_\Delta(V), \quad r \mapsto \overline{r},$$

is a ring homomorphism; the faithfulness of V implies that it is injective. Thus, R is isomorphic to

$$\overline{R} := \{\overline{r} \mid r \in R\} \subseteq \mathrm{End}_\Delta(V).$$

Our goal is to prove that $\overline{R}$ is dense. The desired condition from Definition 5.11 certainly holds for $n = 1$, since $Rv = V$ for every $v \neq 0$ in V.

Let us pause for a moment. So far we have shown that up to isomorphism primitive rings can be characterized as rings satisfying the $n = 1$ case of Definition 5.11. This was obtained almost for free; more precisely, the arguments were straightforward, but, of course, an ingenious recasting of the definition of primitivity was behind them. The second part of the proof is more tricky.

We claim that it is enough to prove the following:

($*$) Let U be a finite dimensional subspace of V. For every $w \in V \setminus U$ there exists $r \in R$ such that $\overline{r}(U) = 0$ and $\overline{r}(w) \neq 0$.

Indeed, assume ($*$) and take linearly independent vectors $u_1, \ldots, u_n \in V$ and arbitrary vectors $v_1, \ldots, v_n \in V$. Given $1 \leq i \leq n$, use ($*$) with the linear span of $u_1, \ldots, u_{i-1}, u_{i+1}, \ldots, u_n$ playing the role of U and u_i playing the role of w. Thus there exists $r_i \in R$ such that $\overline{r_i}(u_j) = 0$ if $j \neq i$ and $\overline{r_i}(u_i) \neq 0$. As we know, we can find $s_i \in R$ such that $\overline{s_i}(\overline{r_i}(u_i)) = v_i$. Note that

$$r := s_1 r_1 + s_2 r_2 + \cdots + s_n r_n \in R$$

satisfies $\overline{r}(u_i) = v_i, i = 1, \ldots, n$.

Suppose ($*$) is not true. Without loss of generality we may assume that U is a space of the smallest dimension for which ($*$) does not hold. Note that $\dim_\Delta U \geq 1$, since ($*$) is obviously true for the zero space. Let U_0 be an arbitrary subspace of U such that $\dim_\Delta U_0 = \dim_\Delta U - 1$. We set

$$L := \{\ell \in R \mid \overline{\ell}(U_0) = 0\}.$$

Take any $z \in V \setminus U_0$. Since ($*$) does hold for U_0 by our assumption, there exists $\ell \in L$ such that $\overline{\ell}(z) \neq 0$. In view of (5.3) this means that $Lz \neq 0$. As L is clearly a left ideal of R, Lz is a submodule of the R-module V. The simplicity of V therefore yields

$$Lz = V \quad \text{for all } z \in V \setminus U_0. \tag{5.4}$$

Let $w \in V \setminus U$ be a vector for which ($*$) does not hold, and choose $u \in U \setminus U_0$. We claim that the map $\delta : V \to V$ given by

$$\delta(\ell u) := \ell w \quad \text{for all } \ell \in L \tag{5.5}$$

is well-defined. First of all, since $Lu = V$ by (5.4), δ is defined on the whole space V. Suppose that $\ell \in L$ is such that $\ell u = 0$, i.e., $\bar{\ell}(u) = 0$. Then $\bar{\ell}$ vanishes on both U_0 and Δu, so that $\bar{\ell}(U) = 0$. Our assumption on w implies that $\bar{\ell}(w) = 0$, i.e., $\ell w = 0$. Therefore δ is indeed well-defined. One easily checks that δ is an R-module endomorphism, i.e., $\delta \in \Delta$. According to (5.2) we can therefore rewrite (5.5) as

$$\ell(\delta u) = \ell w \quad \text{for all } \ell \in L,$$

that is,

$$L(\delta u - w) = 0.$$

From (5.4) it follows that $\delta u - w \in U_0$. However, this yields

$$w = \delta u - (\delta u - w) \in \Delta u + U_0 = U,$$

contradicting the choice of w. This proves (∗).   □

5.3 Alternative Versions and Applications

We have stated the Jacobson Density Theorem in a condensed manner. For applications it is sometimes useful to know that the vector space from the theorem has a tight connection to a faithful simple R-module V. Actually, the faithfulness of V was needed only to assure the injectivity of the canonical homomorphism $r \mapsto \bar{r}$. If V is not faithful, then this homomorphism has a nontrivial kernel, namely $\text{ann}_R(V)$. In this case $R/\text{ann}_R(V)$ is isomorphic to $\bar{R}$. We thus have the following *expanded version of the Jacobson Density Theorem*:

Theorem 5.17 *Let R be a ring and let V be a simple R-module. Consider V as a vector space over the division ring $\Delta = \text{End}_R(V)$. Then the ring $R/\text{ann}_R(V)$ is isomorphic to a dense ring of linear operators of V.*

Our next objective is the *Jacobson Density Theorem for algebras*. Let A be an algebra over a field F. We say that A is a **primitive algebra** if it has a faithful simple A-module V. Here, of course, by a module we mean an algebra module. Therefore $\Delta := \text{End}_A(V)$ is a division F-algebra, not merely a ring. Likewise $\text{End}_\Delta(V)$ is an F-algebra, and $r \mapsto \bar{r}$ is an algebra homomorphism. Accordingly, Theorem 5.16 gets the following form.

Theorem 5.18 *An F-algebra A is primitive if and only if it is isomorphic to a dense algebra of linear operators of a vector space over a division F-algebra.*

We are now in a position to give a different proof of the fundamental Wedderburn's structure theorem (Theorem 2.61).

Corollary 5.19 *Let A be a nonzero finite dimensional algebra. The following statements are equivalent:*

(i)   *A is prime.*
(ii) *A is simple.*
(iii) *There exist a finite dimensional division algebra Δ and a finite dimensional vector space V over Δ such that $A \cong \mathrm{End}_\Delta(V)$ (and hence $A \cong M_n(D)$, where $D = \Delta^\circ$ and $n = \dim_\Delta V$).*
(iv) *A is primitive.*

Proof We know that (iii) implies (ii) (Example 1.10), and trivially (ii) implies (i). Since A is finite dimensional it has a minimal left ideal L. Lemma 5.5, which was stated for rings but obviously it also holds for algebras, therefore tells us that (i) implies (iv). It remains to show that (iv) implies (iii).

Let A be primitive. From Theorem 5.18 we know that A is isomorphic to a dense algebra $\overline{A}$ of linear operators of a vector space V over a division algebra Δ. Let $u_1, \ldots, u_m \in V$ be linearly independent over Δ. Then there exist $f_1, \ldots, f_m \in \overline{A}$ such that $f_i(u_i) \neq 0$ and $f_i(u_j) = 0$ if $j \neq i$. Note that $f_1, \ldots, f_m$ are linearly independent over the base field F. Therefore $m \leq [\overline{A} : F]$. Since $\overline{A}$ is finite dimensional over F, V must be finite dimensional over Δ. Corollary 5.13 implies that $\overline{A} = \mathrm{End}_\Delta(V)$ (and hence $A \cong M_n(D)$, $D = \Delta^\circ$, by Theorem 3.31). Obviously, Δ is also finite dimensional over F. $\square$

We can now somewhat better understand the meaning and importance of the Jacobson Density Theorem. It yields a genuine generalization of Wedderburn's classical theory. This will become even more apparent later, especially when exploring the connection between primitive rings and the Jacobson radical.

Looking back at our proof of Theorem 2.61, it may now seem a bit raw and naive. Why struggle with matrix manipulations when we can obtain the same and more by refined linear operator methods? Note, however, that the arguments in the last proofs are much more complex than those in Chap. 2; juggling with several tools and concepts in one proof may not be so easy for a beginner.

Let us now change the perspective. Suppose we are given a nonzero vector space V over a division F-algebra Δ and an F-subalgebra A of $\mathrm{End}_\Delta(V)$. When is A dense? We can view V as a faithful A-module (via $fv := f(v)$), which is simple if $Av = \{f(v) \mid f \in A\}$ equals V for every nonzero $v \in V$. Let us therefore assume the latter. One should not, however, hastily conclude that this already means that A is dense. We also have to assume that $\mathrm{End}_A(V) = \Delta$ (in this context we identify $\delta \in \Delta$ with the map $V \to V, v \mapsto \delta v$). Obviously, $\Delta \subseteq \mathrm{End}_A(V)$. The converse inclusion does not always hold.

Example 5.20 Consider $\mathbb{C}$ as a 2-dimensional space over $\mathbb{R}$ $(= \Delta = F)$. Let A be the subalgebra of $\mathrm{End}_\mathbb{R}(\mathbb{C})$ consisting of all (left) multiplication maps L_z, $z \in \mathbb{C}$. Obviously, $Av = \mathbb{C}$ for every $0 \neq v \in \mathbb{C}$. It is easy to see that $\mathrm{End}_A(\mathbb{C}) \cong \mathbb{C}$, so it properly contains $\mathbb{R}$. Moreover, from Corollary 5.13 we see that A is not dense.

The *Jacobson Density Theorem for algebras of linear operators* therefore reads as follows.

Theorem 5.21 *Let Δ be a division algebra over a field F, let V be a vector space over Δ, and let A be an F-subalgebra of $\mathrm{End}_\Delta(V)$. If $Av = V$ for every nonzero $v \in V$ and $\mathrm{End}_A(V) = \Delta$, then A is a dense algebra of linear operators of V.*

Remark 5.22 The condition "$Av = V$ for every nonzero $v \in V$" (i.e., the condition that V is simple as an A-module) is equivalent to the condition that there does not exist a proper nonzero subspace W of V that is invariant under every $f \in A$. Indeed, just observe that Av is a subspace invariant under every $f \in A$, and the statement follows.

The next two corollaries are concerned with the simplest situation where $\Delta = F$. The first one is a classical result which was, in a slightly different form connected to the representation of groups, obtained by W. Burnside in 1905.

Corollary 5.23 (Burnside) *Let V be a finite dimensional vector space over an algebraically closed field F. If A is a subalgebra of $\mathrm{End}_F(V)$ such that $Av = V$ for every nonzero $v \in V$, then $A = \mathrm{End}_F(V)$.*

Proof All we have to show is that $\mathrm{End}_A(V) = F$, and this is easy: $\mathrm{End}_A(V)$ is a finite dimensional division algebra over F, so it must be equal to F by Proposition 1.6. $\square$

If a central simple F-algebra A is finite dimensional, then its multiplication algebra $M(A)$ coincides with $\mathrm{End}_F(A)$ (Lemma 1.25). In the next corollary, first proved by E. Artin and G. Whaples a couple of years before Jacobson's discovery of the density theorem, we extend this extremely useful result to infinite dimensional algebras.

Corollary 5.24 (Artin-Whaples) *Let A be a central simple algebra. Then its multiplication algebra $M(A)$ is a dense algebra of linear operators of A.*

Proof Take $0 \neq v \in A$. Since A is simple, the ideal of A generated by v is equal to A. Thus, for every $w \in A$ there exist $a_i, b_i \in A$ such that $\sum_i a_i v b_i = w$. The condition $M(A)v = A$ is thus fulfilled.

It remains to prove that $\mathrm{End}_{M(A)}(A) = F$. That is, we have to show that each $\delta \in \mathrm{End}_{M(A)}(A)$ is a scalar multiple of id_A. To this end we only have to use

$$\delta(L_a(1)) = L_a(\delta(1)) \quad \text{and} \quad \delta(R_a(1)) = R_a(\delta(1)).$$

Indeed, these two formulas can be rewritten as $\delta(a) = a\delta(1) = \delta(1)a$ for every $a \in A$, and therefore $\delta(1) \in F$ since A is central. $\square$

As already remarked after the proof of Theorem 4.42, Corollary 5.24 makes it possible for us to get rid of the assumption that the algebra B from that theorem is finite dimensional.

5.4 Primitive Rings Having Minimal Left Ideals

If the Jacobson Density Theorem says that all primitive rings are, in some sense, close to rings of matrices over division rings, then those with minimal left ideals are closer. This is what we intend to show in this section.

Although we are presently interested in primitive rings, it seems more natural to start things off with general rings.

Lemma 5.25 *Let R be a ring having minimal left ideals. Then the sum of all minimal left ideals of R is a two-sided ideal of R.*

Proof We have to prove that the sum of all minimal left ideals of R is a right ideal. Note that it is enough to show the following: If L is a minimal left ideal of R and $a \in R$ is such that $La \neq 0$, then La is also a minimal left ideal. To this end, consider the map $L \to La$, $x \mapsto xa$. It is obviously a surjective R-module homomorphism. The minimality of L is equivalent to the condition that L, regarded as an R-module, has no proper nonzero submodules. The kernel of our map is therefore equal to 0. Thus, L and La are isomorphic as R-modules. Hence La does not have proper nonzero submodules, meaning that La is a minimal left ideal.   □

Minimal one-sided ideals of a *semiprime* ring R were already considered in Lemmas 2.58 and 2.60. Recall that a left (resp. right) ideal J of R is minimal if and only if J is of the form $J = Re$ (resp. $J = eR$), where e is an idempotent in R such that eRe is a division ring. Furthermore, given an idempotent $e \in R$, Re is a minimal left ideal of R if and only if eR is a minimal right ideal of R. In particular, R has minimal left ideals if and only if R has minimal right ideals.

Lemma 5.26 *Let R be a semiprime ring having minimal one-sided ideals. Then the sum of all minimal left ideals of R coincides with the sum of all minimal right ideals of R.*

Proof Let T be a minimal right ideal of R. As just mentioned, there exists an idempotent $e \in R$ such that $T = eR$ and $L := Re$ is a minimal left ideal of R. Therefore $e = ee \in L$. That is, e lies in a minimal left ideal, and hence $T = eR$ is contained in the sum of all minimal left ideals by Lemma 5.25. Similarly, by using the obvious analogue of Lemma 5.25 for right ideals, we see that every minimal left ideal is contained in the sum of all minimal right ideals.   □

Definition 5.27 Let R be a semiprime ring. If R has minimal one-sided ideals, then let $soc(R)$ denote the sum of all minimal left ideals of R (equivalently, $soc(R)$ is the sum of all minimal right ideals of R). Otherwise we let $soc(R) = 0$. We call $soc(R)$ the **socle** of R.

As $soc(R)$ is a two-sided ideal of R, it is equal to R if R is a simple ring having a minimal one-sided ideal. For example, the ring of matrices over a division ring $R = M_n(D)$ readily coincides with $soc(R)$ since it is equal to the sum of its minimal left ideals RE_{ii}, $i = 1, \ldots, n$ (cf. Example 3.54). However, this example is too plain to reveal the true nature of the socle. The discussion that follows will give a better insight.

From now on we consider the case where R is a *primitive* ring. Let us first recall what we already know. The Jacobson Density Theorem says that there is an isomorphism $r \mapsto \bar{r}$ from R onto a dense ring $\bar{R}$ of linear operators of a vector space V over a division ring Δ. We keep this notation until the end of the section. Assume additionally that R has a minimal left ideal L. Then we can consider L as a faithful simple R-module (in fact, it is enough to assume the primeness of R for this, see Lemma 5.5). Corollary 3.52 tells us that V and Δ can be chosen so that $V = Re$ and $\Delta \cong (eRe)^{\circ}$ for some idempotent $e \in R$. However, we shall not elaborate this view any further. The clue for obtaining additional information about $\bar{R}$ is to consider *all* minimal one-sided ideals of R, not only the chosen minimal left ideal L. More precisely, the socle of R will play a fundamental role.

The following example is supposed to serve as a motivation for the next lemma.

Example 5.28 An example of a minimal left ideal of $R = M_n(D)$ is $L := RE_{11}$, the set of all matrices that have zeros in all columns except in the first one (Example 2.57). By Theorem 3.31 we may consider elements in R as linear operators. Note that all nonzero elements in L have rank one, and each of them generates the left ideal L.

Lemma 5.29 *Let R be a primitive ring, and let $a \in R$. Then the left ideal L generated by a is minimal if and only if the operator $\bar{a}$ has rank one.*

Proof Suppose $\bar{a}$ does not have rank one. We may assume that $a \neq 0$. Therefore there exist $u, v \in V$ such that $\bar{a}(u)$ and $\bar{a}(v)$ are linearly independent over Δ. By density of $\bar{R}$ there exists $r \in R$ such that $\bar{r}(\bar{a}(u)) = 0$ and $\bar{r}(\bar{a}(v)) \neq 0$. Accordingly,

$$I := \{x \in L \mid \bar{x}(u) = 0\}$$

is a left ideal of R which is nonzero (since $0 \neq ra \in I$) and properly contained in L (since $a \notin I$). Therefore L is not minimal.

Assume now that $\bar{a}$ has rank one. Let J be a left ideal of R such that $0 \neq J \subseteq L$. We have to show that $a \in J$. Choose a nonzero $y \in J$, and let $u \in V$ be such that $\bar{y}(u) \neq 0$. By density we can find $r \in R$ satisfying $\bar{r}(\bar{y}(u)) = \bar{a}(u)$. Note that $\bar{a}(u) \neq 0$ since $y \in L = Ra + \mathbb{Z}a$ and $\bar{y}(u) \neq 0$. Now take an arbitrary $v \in V$. Since $\bar{a}$ has rank one, we have $\bar{a}(v) = \delta\bar{a}(u)$ for some $\delta \in \Delta$. Therefore $v - \delta u \in \ker \bar{a} \subseteq \ker \bar{y}$, so that $\bar{y}(v) = \delta\bar{y}(u)$. Consequently,

$$\overline{ry}(v) = \bar{r}(\delta\bar{y}(u)) = \delta\bar{r}(\bar{y}(u)) = \delta\bar{a}(u) = \bar{a}(v).$$

Therefore $a = ry \in J$. □

Describing the socle of a primitive ring is now an easy task.

Theorem 5.30 *Let R be a primitive ring, and let $a \in R$. Then $a \in \text{soc}(R)$ if and only if the operator $\bar{a}$ has finite rank.*

Proof If $a \in \mathrm{soc}(R)$, then $a = a_1 + \cdots + a_n$ where each a_i lies in some minimal left ideal L_i. If $a_i \neq 0$, then $\overline{a_i}$ has rank one by Lemma 5.29. Hence $\overline{a}$ has rank at most n.

Conversely, assume that $\overline{a}$ has finite rank. Let $\{v_1, \ldots, v_m\}$ be a basis of the image of $\overline{a}$. Since $\overline{R}$ is dense, there exist $r_1, \ldots, r_m \in R$ such that $\overline{r_i}(v_i) = v_i$ and $\overline{r_i}(v_j) = 0$ if $j \neq i$. Note that $\overline{r_1} + \cdots + \overline{r_m}$ acts as the identity on the range of $\overline{a}$, and so $a = r_1 a + \cdots + r_m a$. Since each $\overline{r_i a}$ has rank one, $r_i a$ lies in $\mathrm{soc}(R)$ by Lemma 5.29. Consequently, $a \in \mathrm{soc}(R)$. $\qquad\square$

Example 5.31 If $R = \mathrm{End}_\Delta(V)$, then $\mathrm{soc}(R)$ is the set of all finite rank linear operators of V.

We are, of course, interested in the situation where $\mathrm{soc}(R) \neq 0$, i.e., R has minimal left ideals. The next corollaries highlight what Theorem 5.30 says about such rings.

Corollary 5.32 *A primitive ring R has minimal left ideals if and only if $\overline{R}$ contains a nonzero finite rank operator.*

Theorem 5.30 together with Lemma 5.14 gives

Corollary 5.33 *A primitive ring R having minimal left ideals contains a nonzero ideal (in fact, its socle) which is isomorphic to a dense ring of finite rank linear operators of a vector space over a division ring.*

The point here is, of course, that these operators are of finite rank. This brings them close to operators on finite dimensional spaces, and hence close to matrices.

A primitive ring having minimal left ideals thus contains a particularly nice ideal. Do we control the whole ring if we control one of its nonzero ideals? We certainly cannot claim this for general rings. Say, if a ring R is equal to the direct product of two unrelated rings R_1 and R_2, then the properties of the ideal $R_1 \times 0$ have nothing to do with the properties of the other part $0 \times R_2$. However, prime, and therefore primitive rings cannot be decomposed in such a way. Problems on prime rings can be often easily reduced to the same problems on any of their nonzero ideals.

Our final corollary can be considered as an extension of Wedderburn's Theorem 2.61.

Corollary 5.34 *A simple unital ring R having minimal left ideals is isomorphic to $M_n(D)$ for some division ring D and $n \in \mathbb{N}$.*

Proof The simplicity of R implies that $\mathrm{soc}(R) = R$. In particular, $1 \in \mathrm{soc}(R)$, and so $\overline{1}$ has finite rank by Theorem 5.30. Being the unity of a dense ring $\overline{R}$, $\overline{1}$ must be the identity map on V. This means that V is finite dimensional over Δ, and so $\overline{R} = \mathrm{End}_\Delta(V)$ by Corollary 5.13. The desired conclusion now follows from Theorem 3.31. $\qquad\square$

Primitive rings having minimal left ideals form an important, easily approachable, and also somewhat exclusive subclass of primitive rings. Every such ring must in particular contain a nontrivial idempotent, unless it is a division ring. Therefore a primitive domain, which is not a division ring, cannot have minimal left ideals. A concrete example is the Weyl algebra $\mathscr{A}_1$ (cf. Example 2.28). Another important example is a free algebra, which will be introduced in the next chapter.

5.5 Primitive Ideals

Our aim now is to briefly examine the following notion that links primitive rings with the central topic of the second part of this chapter, the Jacobson radical.

Definition 5.35 An ideal P of a ring R is said to be a **primitive ideal** if P is the annihilator of a simple R-module.

(A formally more appropriate term is a *left primitive ideal*—not because it was merely a left ideal, it certainly is two-sided, but because "module" means "left module" by our convention.)

Clearly, R is a primitive ring if and only if 0 is a primitive ideal of R. More generally, we have

Lemma 5.36 *An ideal P of a ring R is a primitive ideal if and only if R/P is a primitive ring.*

Proof Let P be a primitive ideal of R and let M be a simple R-module such that $P = \operatorname{ann}_R(M)$. One immediately checks that the additive group of M becomes a simple R/P-module if we define

$$(r + P)m := rm \text{ for all } r \in R, m \in M.$$

This operation is indeed well-defined since $P \subseteq \operatorname{ann}_R(M)$. On the other hand, from $P \supseteq \operatorname{ann}_R(M)$ we infer that M is a faithful R/P-module. Therefore the ring R/P is primitive.

The proof of the converse is analogous. Starting with a faithful simple R/P-module N, we note that by defining

$$rn := (r + P)n \text{ for all } r \in R, n \in N,$$

the additive group of N becomes a simple R-module (the well-definedness is not an issue here). It is also clear that $\operatorname{ann}_R(N) = P$. Accordingly, P is a primitive ideal. $\square$

Example 5.37 Let $R = R_1 \times \cdots \times R_n$, where each R_i is a primitive ring. Then every

$$P_i = R_1 \times \cdots \times R_{i-1} \times 0 \times R_{i+1} \times \cdots \times R_n \tag{5.6}$$

is a primitive ideal. Indeed, $R/P_i \cong R_i$ is a primitive ring. In the special case where $R_i \cong M_{n_i}(D_i)$ for some $n_i \in \mathbb{N}$ and a division ring D_i, the P_i's are easily seen to be the only primitive ideals of R. Moreover, in this case they are the maximal ideals.

Lemma 5.38 *Let W be an ideal of a ring R. If R/W is a simple ring, then W is a maximal ideal. Conversely, if W is maximal and $R^2 \not\subseteq W$ (this is automatically fulfilled if R is unital), then R/W is a simple ring.*

Proof Let R/W be simple. If J is an ideal of R such that $W \subseteq J \subseteq R$, then J/W is an ideal of R/W, so that $J/W = 0$ or $J/W = R/W$. This readily implies $J = W$ or $J = R$, proving the maximality of W.

If I is an ideal of R/W, then $Z := \{r \in R \mid r + W \in I\}$ is an ideal of R which contains W. Assuming that W is maximal it follows that $Z = W$ or $Z = R$, that is, $I = 0$ or $I = R/W$. Therefore R/W is a simple ring provided that $(R/W)^2 \neq 0$, i.e., $R^2 \nsubseteq W$. $\square$

Since simple unital rings are primitive (Lemma 5.6), Lemmas 5.36 and 5.38 yield

Corollary 5.39 *Every maximal ideal of a unital ring R is a primitive ideal.*

Maximal ideals of $\mathbb{Z}$ and $C[a, b]$, presented in Example 3.44, are therefore examples of primitive ideals. For commutative rings the converse to Corollary 5.39 actually also holds.

Corollary 5.40 *Every primitive ideal of a commutative ring R is a maximal ideal.*

Proof If P is a primitive ideal of R, then R/P is a primitive commutative ring by Lemma 5.36, and hence a field by Lemma 5.7. Therefore P is maximal by Lemma 5.38. $\square$

A *primitive ideal of an algebra A* is, of course, defined as the annihilator of a simple (algebra) A-module. As one would guess, all results from this section still hold if we replace "ring" by "algebra". Since finite dimensional primitive algebras are automatically simple by Corollary 5.19, a similar argument as in the proof of Corollary 5.40 gives

Corollary 5.41 *Every primitive ideal of a finite dimensional algebra A is a maximal ideal.*

In general rings it is easy to find examples of primitive ideals that are not maximal. Say, the zero ideal of a non-simple primitive ring will do (see, e.g., Example 5.10).

Primitive ideals are also related to maximal left ideals.

Lemma 5.42 *If P is a primitive ideal of a ring R, then there exists a maximal left ideal U of R such that $P = \{x \in R \mid xR \subseteq U\}$. Conversely, if U is a maximal left ideal of R and $R^2 \nsubseteq U$ (this is automatically fulfilled if R is unital), then $P := \{x \in R \mid xR \subseteq U\}$ is a primitive ideal of R.*

Proof Let M be a simple R-module such that $P = \text{ann}_R(M)$. By Lemma 3.46 there exists a maximal left ideal U of R such that $M \cong R/U$. Therefore $P = \text{ann}_R(R/U) = \{x \in R \mid xR \subseteq U\}$.

Assume now that U is a maximal left ideal of R such that $R^2 \nsubseteq U$. Lemma 3.46 tells us that R/U is a simple R-module. Consequently, $\text{ann}_R(R/U) = \{x \in R \mid xR \subseteq U\} = P$ is a primitive ideal of R. $\square$

Corollary 5.43 *Every maximal left ideal U of a unital ring R contains a primitive ideal P.*

5.6 Introducing the Jacobson Radical

Recall that the radical of a finite dimensional algebra A is defined as the (necessarily unique) maximal nilpotent ideal of A. A simple-minded attempt to use the same definition for general rings immediately fails, simply because rings may not have maximal nilpotent ideals (Example 2.12). Anyway, generalizations should have some purpose. Our goal is to define the radical of a ring R as an ideal consisting of all the most "undesired" elements in R, that is, the elements whose presence creates an obstacle for understanding the structure of R. Further, we wish the factor ring of R modulo the radical to be "nice" in the sense that its radical is zero, so that we can say something about its structure. As witnessed by Theorem 2.65, the radical of a finite dimensional algebra does satisfy these criteria: the corresponding factor algebra is the direct product of full matrix algebras over division algebras. Of course, we have to be realistic in our expectations when trying to extend this result to general rings. One cannot hope that the structure of "nice" rings is as simple as the structure of "nice" finite dimensional algebras.

There are different radicals defined for general rings, each of them having its own relevance. From the point of view of the structure theory, the following radical is arguably the most important.

Definition 5.44 The **Jacobson radical of a ring** R, denoted by $\mathrm{rad}(R)$, is the intersection of all the primitive ideals of R. If R has no primitive ideals (i.e., R has no simple modules), then we define $\mathrm{rad}(R) = R$.

In other words, $\mathrm{rad}(R)$ consists of all elements that annihilate every simple R-module. It is obviously a (two-sided) ideal of R. Since our modules are actually left modules, apparently a more appropriate name for $\mathrm{rad}(R)$ would be the *left Jacobson radical*. Fortunately, it turns out that the Jacobson radical is a left-right symmetric notion, so this terminological query is unnecessary.

The meaning of the Jacobson radical may not be evident from the definition. In what follows we will discover several properties and characterizations of $\mathrm{rad}(R)$, and thereby give a better insight into its nature. It truly is an amazing concept; we shall see that it touches several seemingly unrelated notions, from simple modules and primitive rings (and hence dense rings of linear operators) to maximal (two-sided and one-sided) ideals, nilpotency and invertibility.

Let us first clarify the relation to nilpotent ideals. Actually, it is not more demanding to consider more general nil ideals.

Lemma 5.45 *Every nil left or right ideal of a ring R is contained in* $\mathrm{rad}(R)$.

Proof Let I be a nil left (resp. right) ideal. If the lemma was not true, there would exist a simple R-module M, $u \in I$ and $m \in M$ such that $n := um \neq 0$. Since $Rn = M$ (Lemma 3.36), we can find $x \in R$ satisfying $xn = m$. That is, $v := xu \in I$ (resp. $v := ux \in I$) satisfies $vm = m$ (resp. $vn = n$), and hence $v^k m = m$ (resp. $v^k n = n$) for every $k \in \mathbb{N}$. But this is impossible for v is nilpotent. $\square$

Later we will see that even a domain may have a large Jacobson radical. Therefore $\mathrm{rad}(R)$ is not always nil, let alone nilpotent.

It goes without saying that the **Jacobson radical of an algebra** A, $\mathrm{rad}(A)$, is defined as the intersection of all primitive (algebra) ideals of A. We may, on the other hand, treat an algebra merely as a ring. Fortunately, the Jacobson radical of the ring A coincides with the Jacobson radical of the algebra A. We suppress the proof of this fact (see, e.g., [Hun74, p. 451]). The only information about the Jacobson radical of an algebra that we need is that it contains all nilpotent ideals. This is clear from the proof of Lemma 5.45.

Theorem 5.46 *If A is a finite dimensional algebra, then* $\mathrm{rad}(A)$ *is the unique maximal nilpotent ideal of A.*

Proof Let N be the unique maximal nilpotent ideal of A (its existence is explained in Sect. 2.1). We know that $N \subseteq \mathrm{rad}(A)$, so we only need to prove the converse inclusion. Take $q \in \mathrm{rad}(A)$. We claim that $q + N \in \mathrm{rad}(A/N)$. If this was not true, there would be a simple A/N-module M such that $(q + N)M \neq 0$. Note that the additive group of M becomes a simple A-module by defining

$$xm := (x + N)m \quad \text{for all } x \in A, m \in M.$$

Since $q \in \mathrm{rad}(A)$, we have $qM = 0$. This contradicts $(q + N)M \neq 0$, and so our claim is proved.

Let us complete the proof by showing that $\mathrm{rad}(A/N) = 0$. By Theorem 2.65 we know that A/N is isomorphic to a finite direct product of full matrix algebras over division algebras. We have thus arrived at the situation considered in Example 5.37. Defining primitive (in fact, maximal) ideals P_i as in (5.6) we clearly have $\bigcap_i P_i = 0$. Consequently, $\mathrm{rad}(A/N) = 0$. $\qquad\square$

The Jacobson radical of a finite dimensional algebra A thus coincides with the radical of A introduced in Sect. 2.1.

5.7   Quasi-invertibility

The Jacobson radical is intimately connected with the concept of invertibility. This statement does not seem to make much sense for non-unital rings. However, we shall rely upon the following notions which make it possible to imitate invertibility in arbitrary rings.

Definition 5.47 An element y in a ring R is said to be **left quasi-invertible** if there exists $x \in R$ such that

$$x + y = xy. \tag{5.7}$$

We call x a **left quasi-inverse** of y. Analogously, $x \in R$ is said to be **right quasi-invertible** if there exists $y \in R$, called a **right quasi-inverse** of x, which satisfies (5.7).

If R is unital, then (5.7) can be written as

$$(1 - x)(1 - y) = 1.$$

Therefore,

x is a left quasi-inverse of y $\iff$ $1 - x$ is a left inverse of $1 - y$.

In particular,

y is left quasi-invertible $\iff$ $1 - y$ is left invertible.

(Thus 1 is not left quasi-invertible.) A similar statement can be made for right quasi-invertibility.

There is another way of looking at these notions. Define $\circ : R \times R \to R$ by

$$x \circ y := x + y - xy.$$

This operation is associative. Indeed, one immediately checks that

$$(x \circ y) \circ z = x + y + z - xy - xz - yz + xyz = x \circ (y \circ z).$$

Clearly $x \circ 0 = 0 \circ x = x$. Accordingly, R endowed with $\circ$ is a monoid with identity element 0. Therefore (5.7) can be interpreted as that x is a left inverse of y with respect to $\circ$. Since the left and right inverses in a monoid coincide, $x \circ y = y \circ z = 0$ implies $x = z$. We can reword this as follows.

Lemma 5.48 *If an element y has a left quasi-inverse x and a right quasi-inverse z, then $x = z$.*

Definition 5.49 An element $y \in R$ is said to be **quasi-invertible** if there exists $x \in R$ which is both a left and right quasi-inverse of y. In this case we call x a **quasi-inverse** of y.

In other words, x is a quasi-inverse of y if it satisfies (5.7) and $xy = yx$. Actually, x is *the* quasi-inverse of y; its uniqueness follows from Lemma 5.48. We also see that $y \in R$ is quasi-invertible if and only if it is both left and right quasi-invertible. If R is unital, this is further equivalent to the invertibility of $1 - y$.

Example 5.50 Every nilpotent element $y \in R$ is quasi-invertible. Indeed, $-\sum_{n=1}^{\infty} y^n$ is its quasi-inverse (the nilpotency of y implies that this sum is actually finite). If R is unital, we can rewrite this observation in the familiar form: $(1 - y)^{-1} = \sum_{n=0}^{\infty} y^n$.

Definition 5.51 We say that $S \subseteq R$ is a **quasi-invertible set** if all elements in S are quasi-invertible. Analogously we define a **left** (resp. **right**) **quasi-invertible set**.

Lemma 5.52 *If a left ideal L is left quasi-invertible, then L is quasi-invertible.*

Proof Take $\ell \in L$. By assumption ℓ has a left quasi-inverse $k \in R$. From $k = k\ell - \ell$ we see that k also lies in L. Thus, k has a left quasi-inverse m. Lemma 5.48 implies $\ell = m$, meaning that k is the quasi-inverse of ℓ. $\qquad\square$

This simple lemma will eventually lead to the left-right symmetry of the Jacobson radical.

Assume for a moment that R is unital. If $1 - a \in R$ is not left invertible, then $R(1 - a)$ is a proper left ideal of R, so it is contained in some maximal left ideal U (Lemma 3.45), which gives rise to a simple R-module R/U (Lemma 3.46). These observations can be easily extended to the non-unital situation.

Lemma 5.53 *Let R be an arbitrary ring. If $a \in R$ is not left quasi-invertible, then there exists a maximal left ideal U of R such that $a \notin U$. Moreover, R/U is a simple R-module and $a \notin \operatorname{ann}_R(R/U)$.*

Proof The assumption that a is not left quasi-invertible can be read as that a does not lie in the left ideal $L := \{ra - r \mid r \in R\}$. Moreover, we claim that if L' is a proper left ideal containing L, then $a \notin L'$. Indeed, by writing $r \in R$ as $r = -(ra - r) + ra$ we see that R is the only left ideal of R that contains $L \cup \{a\}$. This makes it possible for us to apply Zorn's lemma in a similar way as in the proof of Lemma 3.45. Let $\mathscr{S}$ be the set of all proper left ideals of R that contain L. Partially order $\mathscr{S}$ by inclusion. If $\{K_i \mid i \in I\}$ is a chain in $\mathscr{S}$, then $K = \bigcup_{i \in I} K_i$ is its upper bound. Indeed, K is a proper left ideal since $a \notin K$. Therefore $\mathscr{S}$ has a maximal element U. Clearly, U is a maximal left ideal and $a \notin U$. Furthermore, since $a - a^2 \in L \subseteq U$ it follows that $a^2 \notin U$. Lemma 3.46 now tells us that R/U is a simple R-module. Finally, $a(a + U) \neq 0$ shows that $a \notin \operatorname{ann}_R(R/U)$. $\qquad\square$

This lemma in particular shows that for every element that is not left quasi-invertible we can find a simple module which is not annihilated by this element. Elements from $\operatorname{rad}(R)$ are therefore necessarily left quasi-invertible, and hence quasi-invertible by Lemma 5.52. This observation can be sharpened as follows:

Theorem 5.54 *Let R be a ring and let $q \in R$. The following statements are equivalent:*

(i) *The ideal generated by q is quasi-invertible.*
(ii) *The left ideal Rq is quasi-invertible.*
(iii) $q \in \operatorname{rad}(R)$.

Proof (i) $\Longrightarrow$ (ii). Trivial.

(ii) $\Longrightarrow$ (iii). Suppose $q \notin \operatorname{rad}(R)$. Then there exists a simple R-module M such that $qm \neq 0$ for some $m \in M$. Accordingly, $Rqm = M$ by Lemma 3.36. Therefore

$\ell m = m$ for some $\ell \in Rq$. Since ℓ is quasi-invertible, there exists $r \in R$ such that $r + \ell = r\ell$. Consequently, $m = \ell m = r\ell m - rm = 0$, which contradicts $qm \neq 0$.

(iii) $\Longrightarrow$ (i). As noticed before the statement of the theorem, Lemmas 5.52 and 5.53 show that $\mathrm{rad}(R)$ is a quasi-invertible ideal. Its subsets are therefore also quasi-invertible. $\qquad\square$

By making obvious modifications in the proof, one shows that (i) is also equivalent to (ii') the *right* ideal qR is quasi-invertible, and (iii') q annihilates all simple right R-modules. We have thereby established that the *left* Jacobson radical coincides with the *right* Jacobson radical (see the comment following Definition 5.44).

The following corollary is immediate.

Corollary 5.55 *The Jacobson radical of a ring R is a quasi-invertible ideal which contains every quasi-invertible one-sided ideal of R.*

Let us mention that Corollary 5.55 implies Lemma 5.45 since nilpotent elements are quasi-invertible (see Example 5.50).

Corollary 5.56 *The Jacobson radical of a unital ring R is equal to the intersection of all the maximal left ideals of R.*

Proof Let T denote the intersection of all the maximal left ideals of R. Clearly, T is a left ideal. Lemma 5.53 along with Lemma 5.52 implies that T is quasi-invertible. Hence $T \subseteq \mathrm{rad}(R)$ by Corollary 5.55. The fact that $\mathrm{rad}(R) \subseteq T$ clearly follows from Corollary 5.43. $\qquad\square$

It is interesting in its own right that the intersection of all the maximal left ideals is a two-sided ideal. Of course we may replace "left" by "right" in the statement of Corollary 5.56.

For the reader's convenience we gather together various characterizations of the Jacobson radical of a unital ring in the next corollary.

Corollary 5.57 *Let R be a nonzero unital ring. The following statements are equivalent for $q \in R$:*

(i) $q \in \mathrm{rad}(R)$.
(ii) $qM = 0$ *for every simple R-module M.*
(iii) q *lies in every ideal P of R such that R/P is a primitive ring.*
(iv) $1 - xq$ *is invertible for all $x \in R$.*
(v) $1 - \sum_i x_i q y_i$ *is invertible for all $x_i, y_i \in R$.*
(vi) q *lies in every maximal left ideal of R.*

One can add appropriate right versions of (ii)–(iv) and (vi).

5.8 Computing the Jacobson Radical

After discovering a variety of abstract descriptions of the Jacobson radical, it may still be unclear how to determine $\text{rad}(R)$ for a given ring R. There is no magical device for this, but we can gain intuition by examining some special situations.

Let us begin with two illustrative examples. Since we know that the Jacobson radical of a finite dimensional algebra is the maximal nilpotent ideal (Theorem 5.46), our first example is just a repetition of Example 2.14.

Example 5.58 The Jacobson radical of $T_n(F)$, the algebra of all upper triangular $n \times n$ matrices over a field F, is the set of all strictly upper triangular matrices.

The ring in the next example has no nonzero nilpotent elements, in fact it is a domain. Yet its Jacobson radical is quite big.

Example 5.59 Take a division ring D and consider the ring $D[[\omega]]$ of formal power series. Let T denote its ideal generated by ω, i.e., T is the set of all elements with zero constant term. Lemma 1.40 says that $f \in D[[\omega]]$ is not invertible if and only if $f \in T$. This readily implies that every proper one-sided ideal of $D[[\omega]]$ is contained in T. Therefore T is the unique maximal left ideal of $D[[\omega]]$. Corollary 5.56 shows that $\text{rad}(D[[\omega]]) = T$.

The Jacobson radical is compatible with various ring constructions. Let us consider only the simplest construction, i.e., the direct product.

Proposition 5.60 *Let* $\{R_i \mid i \in I\}$ *be a family of rings. Then*

$$\text{rad}(\Pi_{i \in I} R_i) = \Pi_{i \in I}\text{rad}(R_i).$$

Proof Note that (r_i) is quasi-invertible in $R := \Pi_{i \in I} R_i$ if and only if r_i is quasi-invertible in R_i for every i. Accordingly, $q = (q_i) \in R$ is such that Rq is quasi-invertible in R if and only $R_i q_i$ is quasi-invertible in R_i for every $i \in I$. Thus, by Theorem 5.54, $q \in \text{rad}(R)$ if and only if $q_i \in \text{rad}(R_i)$ for every $i \in I$. $\square$

Relating the Jacobson radical of a ring to the Jacobson radical of some of its subrings is another topic of interest. For ideals the relation is simple and nice.

Proposition 5.61 *If I is an ideal of a ring R, then* $\text{rad}(I) = I \cap \text{rad}(R)$.

Proof Let $y \in I \cap \text{rad}(R)$. As an element of $\text{rad}(R)$, y is quasi-invertible in R. If x is its quasi-inverse, then we see from $x = xy - y$ that x also lies in I. This shows that $I \cap \text{rad}(R)$ is a quasi-invertible ideal of the ring I. Therefore $I \cap \text{rad}(R) \subseteq \text{rad}(I)$ by Corollary 5.55.

Proving the converse inclusion is a bit more tricky. Take $q \in \text{rad}(I)$ and $r \in R$. In view of Theorem 5.54 it is enough to show that $u := rq$ is quasi-invertible in R. Since $u^2 = (rqr)q \in I\text{rad}(I) \subseteq \text{rad}(I)$, it follows that u^2 is quasi-invertible. We claim that this implies that u is also quasi-invertible. In unital rings this follows easily from $1 - u^2 = (1 - u)(1 + u)$, but of course we do not wish to assume the existence of unity. At any rate, the proof is easy. A direct computation shows that if v is the quasi-inverse of u^2, then $vu + v - u$ is a left quasi-inverse of u and $uv + v - u$ is a right quasi-inverse of u. Thus u is quasi-invertible (and $uv = vu$) by Lemma 5.48. □

If the size of rad(R) measures the intricacy of the structure of R, the following definition considers the worse (yet not the least interesting) situation.

Definition 5.62 A ring R is said to be a **radical ring** if rad(R) = R.

Let us mention two alternative ways to state this definition.

Proposition 5.63 *The following statements are equivalent for a ring R:*

 (i) *R is a radical ring.*
 (ii) *Every element in R is quasi-invertible.*
(iii) *R has no simple modules.*

Proof From Corollary 5.55 we see that (i) and (ii) are equivalent. If R is a radical ring, then R cannot have simple modules since their annihilators are always proper subsets of R. Thus, (i) implies (iii), while the converse holds by Definition 5.44.

Proposition 5.64 *The Jacobson radical of an arbitrary ring R is a radical ring, i.e.,* rad(rad(R)) = rad(R).

Proof Take rad(R) for I in Proposition 5.61. □

Examples 5.58 and 5.59 therefore also yield examples of radical rings. The one from the first example, the ring of all strictly upper triangular matrices, is nilpotent. Such rings are automatically radical; more generally, every nil ring is a radical ring by Lemma 5.45.

5.9 Semiprimitive Rings

After considering the most unfavorable situation where rad(R) = R, let us now examine the opposite extreme.

Definition 5.65 A ring R is said to be a **semiprimitive ring** if rad(R) = 0.

Example 5.66 Every primitive ring is semiprimitive since 0 is its primitive ideal. More generally, the direct product of primitive rings (in fact, even of semiprimitive rings) is semiprimitive by Proposition 5.60.

Example 5.67 If R is a nonzero commutative unital ring, then rad(R) is equal to the intersection of all the maximal ideals of R by Corollary 5.56. With reference to Example 3.44 we thus have:

(a) Since the intersection of all the maximal ideals $p\mathbb{Z}$ of $\mathbb{Z}$ is 0 (here, p is a prime number), $\mathbb{Z}$ is a semiprimitive ring.
(b) Similarly, since $\bigcap_{a \le c \le b} U_c = 0$, where $U_c = \{f \in C[a, b] \mid f(c) = 0\}$, $C[a, b]$ is a semiprimitive ring.

A semiprimitive ring R cannot have nonzero nil ideals by Lemma 5.45. In particular, semiprimitive rings are thus semiprime. The converse is not true.

Example 5.68 Let D be a division ring. In Example 5.59 we saw that $D[[\omega]]$ has a big Jacobson radical. Thus, $D[[\omega]]$ is an example of a semiprime ring (in fact, a domain) that is not semiprimitive. The polynomial ring $D[\omega]$, however, is semiprimitive. Indeed, take $f \in \text{rad}(D[\omega])$. Then f is quasi-invertible by Corollary 5.55; that is, $1 - f$ is invertible. Since polynomials of degree 0 are clearly the only invertible elements in $D[\omega]$, and since $f \in \text{rad}(D[\omega])$ cannot be invertible, it follows that $f = 0$.

The reader might have noticed that we did not use the full force of the assumption that D is a division ring. Instead of analyzing what else the above simple argument brings, we mention a much deeper result by S. A. Amitsur from 1956: Every ring R has a nil ideal N such that $\text{rad}(R[\omega]) = N[\omega]$. Consequently, $R[\omega]$ is semiprimitive if R has no nonzero nil ideal. But even Amitsur's theorem is not the end of the story. A precise description of N is still a mystery. It turns out that the validity of the statement "*If R is nil, then $N = R$*" is equivalent to a positive answer to Köthe's problem.

It is not always easy to find out whether or not a ring is semiprimitive. Here is a famous old problem:

Problem 5.69 Let G be a group and let F be a field with $\text{char}(F) = 0$. Is the group algebra $F[G]$ semiprimitive?

The answer is "yes" if G is finite. Indeed, this is just a rewording of Maschke's theorem (namely, as is evident from Theorem 5.46, a finite dimensional algebra is semiprime if and only if it is semiprimitive). Thus, we are asking whether Maschke's theorem can be generalized to infinite groups. It turns out that the problem can be reduced to the case where $F = \mathbb{Q}$.

Let us return to problems we can handle. At the beginning of Sect. 5.6 we said that a decent radical should have the property that the factor ring of R modulo the radical has zero radical. We still owe the proof that the Jacobson radical does satisfy this property.

Lemma 5.70 *The ring $R/\text{rad}(R)$ is semiprimitive for every ring R.*

Proof The case where R is a radical ring is trivial. By Proposition 5.63, we may therefore assume that R has simple modules. Take such a module M. The additive group of M becomes a simple module over $R/\text{rad}(R)$ by defining

$$(x + \text{rad}(R))m := xm \quad \text{for all } x \in R, m \in M.$$

Checking this is straightforward, including the well-definedness which follows from $\mathrm{rad}(R)M = 0$. Now, if $q + \mathrm{rad}(R) \in \mathrm{rad}(R/\mathrm{rad}(R))$, then $qM = (q + \mathrm{rad}(R))M = 0$. Since M is an arbitrary simple R-module, this gives $q \in \mathrm{rad}(R)$. □

Another debt to the reader is an explanation why semiprimitive rings are supposed to be "nice". Why would it be easier to deal with a semiprimitive ring than with any old ring? For primitive rings this is clear as we have the Jacobson Density Theorem. Direct products of primitive rings can therefore also be viewed as easily approachable. Unfortunately, not every semiprimitive ring can be represented as such a product (say, $\mathbb{Z}$ cannot). We are forced to deal with the following generalization of the direct product.

Definition 5.71 Let $\{R_i \mid i \in I\}$ be a family of rings. A subring R of the direct product $\Pi_{i \in I} R_i$ is called a **subdirect product** of the family $\{R_i \mid i \in I\}$ if for each $j \in I$, the map $\pi_j : R \to R_j$, $\pi_j((r_i)) = r_j$, is surjective.

If $I = \{1, 2\}$, this means that for every $r_1 \in R_1$ there exists $r_2 \in R_2$ such that $(r_1, r_2) \in R$, and similarly, for every $r_2 \in R_2$ there exists $r_1 \in R_1$ such that $(r_1, r_2) \in R$. For instance, every ring R is isomorphic to the subdirect product of two copies of R via $r \mapsto (r, r)$. Subdirect products are thus quite different from direct products.

Lemma 5.72 *A nonzero ring R is semiprimitive if and only if R is isomorphic to a subdirect product of primitive rings.*

Proof Let R be semiprimitive. Denote the family of all the primitive ideals of R by $\{P_i \mid i \in I\}$. Recall that the rings R/P_i are primitive (Lemma 5.36). By assumption, $\bigcap_{i \in I} P_i = 0$. This shows that the homomorphism

$$R \to \Pi_{i \in I} R/P_i, \quad r \mapsto (r + P_i) \tag{5.8}$$

is injective. Its image is clearly a subdirect product of the family $\{R/P_i \mid i \in I\}$.

Conversely, take a family of primitive rings $\{R_i \mid i \in I\}$ and let $\varphi : R \to \Pi_{i \in I} R_i$ be an injective homomorphism such that $\pi_j(\varphi(R)) = R_j$ for every $j \in I$. Set $P_j := \ker \pi_j \varphi$. Since $R/P_j \cong R_j$ we see, again by Lemma 5.36, that P_j is a primitive ideal of R. If $r \in \bigcap_{i \in I} P_i$, then $\varphi(r) = 0$ and hence $r = 0$. Thus $\mathrm{rad}(R) \subseteq \bigcap_{i \in I} P_i = 0$. □

Example 5.73 Let $R = A$ be a finite dimensional semiprimitive algebra. As mentioned above, A is then in particular semiprime, and hence, by Wedderburn's theorem, isomorphic to a finite direct product of simple (= primitive in this context) algebras. Therefore A has only finitely many primitive ideals (cf. Example 5.37). Note that the homomorphism from (5.8) is surjective.

Example 5.74 From Example 5.67 and the proof of Lemma 5.72 we infer the following:

(a) $\mathbb{Z}$ is isomorphic to a subdirect product of the fields $\mathbb{Z}_p$, p prime.
(b) $C[a,b]$ is isomorphic to a subdirect product of the fields $C[a,b]/U_c \cong \mathbb{R}$, $a \le c \le b$ (cf. Example 3.44).

Roughly speaking, Lemma 5.72 shows that a semiprimitive ring can be decomposed into primitive factors, however, in a fashion which is not always easily controllable. When making this decomposition we somehow lose track of the way the components are "glued together". The example with $C[a,b]$ is illustrative. The direct product of the fields $C[a,b]/U_c \cong \mathbb{R}$, $a \le c \le b$, can be identified with the ring of all functions from $[a,b]$ into $\mathbb{R}$. Thus, when applying the method of Lemma 5.72 to $C[a,b]$ we lose the information that functions from our ring are continuous. The structure theory is powerful, but has its limitations.

Our final result in this section summarizes some of the main features of the structure theory. One can compare it with the analogous result concerning finite dimensional algebras, Theorem 2.65.

Theorem 5.75 *If a ring R is not a radical ring, then the ring $R/\mathrm{rad}(R)$ is isomorphic to a subdirect product of dense rings of linear operators of vector spaces over division rings.*

Proof Combine Lemmas 5.70 and 5.72 with the Jacobson Density Theorem. □

5.10 Structure Theory in Action

The structure theory that we have developed is not only beautiful, it is truly applicable. The purpose of this final section is to indicate how the theory works in "real life".

Suppose we are given a problem on a ring R. The following procedure, indicated by Theorem 5.75, sometimes leads to its solution:

(a) By factoring out $\mathrm{rad}(R)$ one reduces the problem to the case where R is semiprimitive.
(b) Using the representation of a semiprimitive ring R as a subdirect product of primitive rings, one further reduces the problem to the case where R is primitive. (In practice one does not necessarily work with subdirect products, but uses the fact that the intersection of the primitive ideals of R is 0. We have introduced subdirect products primarily to indicate the philosophical aspect of semiprimitivity.)
(c) A problem on a primitive ring R can be usually solved by representing R as a dense ring of linear operators of a vector space V over a division ring Δ.
(d) If $\dim_\Delta V = 1$, the density condition does not have the same meaning as otherwise. In this case R is a division ring (isomorphic to Δ°) and a different argument may be needed.

Needless to say, we cannot solve just any problem by following this procedure. There are, however, various important instances where at least some, if not all of the steps (a)–(d) can be realized. We will give an evidence of the feasibility of the procedure by presenting a proof of (yet another) theorem by Jacobson. This theorem was one of the first applications of the structure theory, and still is one of the most striking ones.

We first record two elementary lemmas. The reader is probably familiar with the first one. Anyway, we give the proof for the sake of completeness.

Lemma 5.76 *If K is a finite field with $\mathrm{char}(K) = p$, then there exists $n \in \mathbb{N}$ such that $|K| = p^n$ and $k^{p^n} = k$ for every $k \in K$. Moreover, if $k_1, \ldots, k_{p^n-1}$ are the nonzero elements in K, then*

$$\omega^{p^n} - \omega = (\omega - k_1)\ldots(\omega - k_{p^n-1})\omega. \tag{5.9}$$

Proof We may consider K as a vector space over the prime subfield $F_0 \cong \mathbb{Z}_p$. If K is n-dimensional over F_0, then $|K| = |F_0^n| = p^n$. The multiplicative group $K^* = K\setminus\{0\}$ thus has $p^n - 1$ elements, so that $k^{p^n-1} = 1$ for every $k \in K^*$. Consequently, $k^{p^n} = k$ for every $k \in K$. That is to say, every element in K is a root of the polynomial $\omega^{p^n} - \omega$. Since the degree of this polynomial is equal to $|K|$, the desired conclusion follows.   □

The next lemma concerns powers of inner derivations in rings of prime characteristic. We denote the inner derivation $x \mapsto ax - xa$ by $\mathrm{ad}\, a$ (this notation is borrowed from the theory of Lie algebras). We first remark that in any ring R, regardless of its characteristic, for all $a, x \in R$ we have

$$(\mathrm{ad}\, a)^m(x) = \sum_{k=0}^m (-1)^k \binom{m}{k} a^{m-k} x a^k. \tag{5.10}$$

Using $\binom{m}{k-1} + \binom{m}{k} = \binom{m+1}{k}$ one easily proves this by induction on m.

Lemma 5.77 *If R is a ring with $\mathrm{char}(R) = p$, where p is prime, then $(\mathrm{ad}\, a)^{p^n} = \mathrm{ad}\, a^{p^n}$ for all $a \in R$ and $n \in \mathbb{N}$.*

Proof If $1 \leq k \leq p-1$, then p does not divide $k!(p-k)!$, so it divides $\binom{p}{k} = \frac{p!}{k!(p-k)!}$. Therefore (5.10) reduces to $(\mathrm{ad}\, a)^p = \mathrm{ad}\, a^p$ if $m = p$ (if $p = 2$ then this is true as $xa^2 = -xa^2$). This proves the claim of the lemma for $n = 1$. We may therefore assume the claim is true for $n - 1$. Consequently,

$$(\mathrm{ad}\, a)^{p^n} = \left((\mathrm{ad}\, a)^{p^{n-1}}\right)^p = \left(\mathrm{ad}\, a^{p^{n-1}}\right)^p = \mathrm{ad}\,\left(a^{p^{n-1}}\right)^p = \mathrm{ad}\, a^{p^n},$$

as desired.   □

Let us now tackle the Jacobson's theorem. First, a few words about the background. The reader is perhaps familiar with the notion of a *Boolean ring*. This is a ring in which every element is an idempotent (e.g., a direct product of copies of $\mathbb{Z}_2$). One of the first things one learns about such rings is that they are commutative. The proof is easy: If x and y are elements of a Boolean ring R, then $x, y, x + y$ are idempotents, which readily implies $xy + yx = 0$; setting $y = x$ we get char$(R) = 2$, so $xy + yx = 0$ can be written as $xy = yx$. This was extremely easy, but the fact itself may appear a bit surprising at first glance since idempotents in rings do not always commute. Now consider a more general condition where a ring R satisfies $x^3 = x$ for every $x \in R$. Is R commutative? It turns out that this is true indeed, but giving a proof may already be a small challenge for the reader. The case where $x^4 = x$ is yet more entangled, but again the commutativity can be derived by ad hoc arguments. If we wish to consider the general case where there exists $n \geq 2$ such that

$$x^n = x \quad \text{for all } x \in R, \tag{5.11}$$

the need for a conceptual approach becomes apparent. But first of all, why would anyone be interested in (5.11)? As noticed in Lemma 5.76, every finite field F satisfies (5.11) with $n = |F|$. The very same proof shows that the same holds for every finite division ring. Surely, such rings are automatically fields by Wedderburn's theorem (Theorem 1.38), but this is exactly the point we wish to make. We are wondering whether Wedderburn's theorem can be generalized to arbitrary rings via (5.11). And indeed it can. Even better, there is no need to fix n; it may depend on x.

Theorem 5.78 (Jacobson) *If a ring R is such that for every $x \in R$ there exists an integer $n(x) \geq 2$ such that $x^{n(x)} = x$, then R is commutative.*

Proof We will follow the procedure outlined above.

(a) In the first step we will show that rad$(R) = 0$ (so that there is no need to factor out rad(R)). Choose $q \in$ rad(R), and set $n = n(q)$. Then q^{n-1} also lies in rad(R), so it is quasi-invertible (Corollary 5.55). Thus, there exists $r \in R$ such that $q^{n-1} + r = rq^{n-1}$. Multiplying this identity from the right by q and using $q^n = q$ we obtain $q = 0$.

(b) We now know that R is semiprimitive. We may assume that $R \neq 0$, so that there is a family $\{P_i \mid i \in I\}$ of primitive ideals of R such that $\bigcap_{i \in I} P_i = 0$. Therefore R is commutative if and only if each primitive ring R/P_i is commutative. Since R/P_i satisfies the same condition, $x^{n(x)} = x$ for every $x \in R/P_i$, we see that without loss of generality we may assume that R itself is primitive.

(c) In view of the Jacobson Density Theorem, we may assume that R is a dense ring of linear operators of a vector space V over a division ring Δ. Suppose there exist linearly independent vectors $u_1, u_2 \in V$. Choose $f \in R$ such that $f(u_1) = u_2$ and $f(u_2) = 0$. Then $f^n(u_1) = 0$ for every $n \geq 2$, while $f(u_1) \neq 0$. However, this contradicts our assumption that $f^{n(f)} = f$. The result is thus proved if $\dim_\Delta V \geq 2$.

It remains to treat the case where $\dim_\Delta V = 1$, i.e., the case where $R \cong \Delta^\circ$ is a division ring. This is the difficult part of the proof.

(d) Let us write D instead of R. We begin by pointing out two facts that follow immediately from the condition $x^{n(x)} = x$, $x \in D$. The first one is that every nonzero subring of D is automatically a division ring (since the inverse of x is a power of x), and the second one is that D has finite characteristic p (indeed, just take $2 = 1 + 1$ for x).

Suppose, on the contrary, that D is not commutative. Take $a \in D \backslash Z(D)$. Since D has finite characteristic and $a^{n(a)-1} = 1$, we see that the subring generated by a, let us call it K, is finite. As observed in the previous paragraph, K is necessarily a field, so we can invoke Lemma 5.76. Thus, there exists $n \in \mathbb{N}$ such that $|K| = p^n$ and, in particular, $a^{p^n} = a$. Consider D as a vector space over K. The inner derivation $d := \mathrm{ad}\, a$ is clearly a K-linear map, so it can be considered as an element of the K-algebra $\mathrm{End}_K(D)$. We can therefore evaluate the polynomial from (5.9) at d. Since $d^{p^n} = d$ by Lemma 5.77, we obtain

$$(d - k_1) \ldots (d - k_{p^n - 1})d = 0, \tag{5.12}$$

where the k_i's are the nonzero elements of K. Note that, in this context, k_i should be understood as the map $D \to D$, $x \mapsto k_i x$. Now, since $a \notin Z(D)$, d is a nonzero map. From (5.12) it thus follows that at least one of the maps $d - k_i$ has a nontrivial kernel. This means that there is $b \in D$ such that $ab - ba = d(b) = k_i b$ with $0 \neq k_i \in K$. Accordingly, a and b do not commute and $ba \in Kb$.

Let D_0 be the subring of D generated by a and b. As observed at the beginning, D_0 is in fact a division ring. Of course, it is noncommutative for $ab \neq ba$. From $ba \in Kb$ it clearly follows that $ba^r \subseteq Kb$ for every $r \in \mathbb{N}$. Hence $bK \subseteq Kb$, and consequently, $b^s K \subseteq Kb^s$ for every $s \in \mathbb{N}$. This shows that

$$K + Kb + \cdots + Kb^{n(b)-2}$$

is a subring of D. Since it is a subset of D_0 and since it contains a and b, it is actually equal to D_0. But then D_0 is a finite set. However, this contradicts Wedderburn's theorem on finite division rings (Theorem 1.38). $\square$

The first three steps (a)–(c) were quite easy, but nevertheless fascinating. They illustrated how the structure theory can be used to reduce a problem on a general ring to a much simpler ring, in this case a division ring. After that we were left to ourselves, and so the last step (d) had to be more ingenious.

It should be mentioned that Jacobson's original proof was somewhat different. We have mainly followed the proof due to I. N. Herstein.

Exercises

5.1. Let e be a nonzero idempotent in a ring R. Show that eV is a simple eRe-module if V is a simple R-module and $eV \neq 0$. Hence infer that eRe is primitive if R is primitive. Use this to show that the rings $M_n(\mathbb{Z})$ and $M_n(F[\omega])$ are prime but not primitive.

5.2. Show that the ring $M_n(R)$ is primitive if and only if R is primitive.

5.3. Let R be a primitive ring with a faithful simple R-module V. Show that R is a division ring if and only if for all $r \in R$ and $v \in V$, $rv = 0$ implies $r = 0$ or $v = 0$.

5.4. Let R be a primitive ring. Suppose that $a, b \in R$ satisfy $(xa)^2 = (xb)^2$ for every $x \in R$. Show that either $a = b$ or $a = -b$.

5.5. Give an example showing that the assumption that F is algebraically closed in Burnside's theorem (Corollary 5.23) is necessary.

5.6. Give an alternative proof of Lemma 1.24 by using the Artin-Whaples Theorem (Corollary 5.24).

5.7. Show that the Weyl algebra $\mathscr{A}_1$ is a dense algebra of linear operators of $F[\omega]$.

   Remark: Since $\mathscr{A}_1$ is primitive (Example 5.9), the Jacobson Density Theorem for algebras tells us that $\mathscr{A}_1$ is isomorphic to a dense algebra of linear operators of a vector space over some division algebra. What has to be shown is that $\mathscr{A}_1$ is a dense algebra of linear operators of the original space $F[\omega]$, i.e., the fulfillment of the conditions of Theorem 5.21 has to be verified.

5.8. Let R be a ring, V be an R-module, and φ be an automorphism of R. Note that the additive group of V becomes an R-module if we define the module operation by $x \cdot v = \varphi(x)v$ for all $x \in R$, $v \in V$. If V is faithful and simple, then so is the new module. Use this construction to show that the Weyl algebra $\mathscr{A}_1$ has nonisomorphic faithful simple modules.

5.9. Let R be a primitive ring having a minimal left ideal L. Show that every faithful simple R-module is isomorphic to L.

   Remark: This is a generalization of Lemma 3.40.

5.10. Fields are the only commutative primitive rings (Lemma 5.7), and the center of a simple unital ring is necessarily a field (Example 1.21). However, the center of a unital primitive ring may not be a field. Find an example of such a ring.

   Hint: If R is a primitive ring having minimal left ideals, then every subring of R that contains $\mathrm{soc}(R)$ is also primitive.

5.11. Show that the sum of minimal left ideals of the upper triangular matrix ring $T_2(F)$ does not coincide with the sum of its minimal right ideals.

5.12. Let R be a primitive ring having minimal left ideals. Show that $\mathrm{soc}(R)$ is simple as a ring, and is contained in every nonzero ideal of R.

5.13. Let R be a unital ring, and let M be a nonzero finitely generated unital R-module. Show that $\mathrm{rad}(R)M$ is a proper submodule of M.

   Remark: This result, known as **Nakayama's lemma**, is particularly well-known in commutative algebra and algebraic geometry.

5.14. A unital ring R is said to be **local** if the set of all non-invertible elements in R forms an ideal. Show that this ideal is exactly $\mathrm{rad}(R)$, and that R is local if and only if $R/\mathrm{rad}(R)$ is a division ring. Verify that the following rings are local:

    (a) The ring $D[[\omega]]$ where D is a division ring (cf. Example 5.59).

    (b) The subring of $M_n(D)$, D a division ring, consisting of upper triangular matrices all of whose diagonal entries are equal.

    (c) $\mathbb{Z}_{p^n}$ where p is prime and $n \in \mathbb{N}$.

    (d) The unitization of a radical algebra.

5.15. Let a be an element in a unital ring R. Show that the following statements are equivalent:

    (i) a is invertible.

    (ii) $a + \mathrm{rad}(R)$ is invertible in $R/\mathrm{rad}(R)$.

    (iii) $a + P$ is invertible in R/P for every primitive ideal P of R.

Hint: If a does not have a left inverse, then there exists a maximal left ideal containing a. If a has a left inverse b, then $(1 - ab)a = 0$.

5.16. Let a be an element in a ring R. Show that if there exists $n \in \mathbb{N}$ such that a^n is quasi-invertible, then so is a.

Hint: After examining the (intuitively more clear) case where R is unital, one can easily guess how to handle the general case.

5.17. Let a be a nonzero element in a ring R such that $a^n = a$ for some $n \geq 2$. Show that $a \notin \mathrm{rad}(R)$. (In particular, $\mathrm{rad}(R)$ does not contain nonzero idempotents.)

5.18. Let a, b be elements in a ring R. Show that aRb is a quasi-invertible set if and only if $ba \in \mathrm{rad}(R)$.

Hint: This can be proved either directly (it is not difficult) or as an application of the theory along with the following (quite remarkable) fact: For any $u, v \in R$, uv is quasi-invertible if and only if vu is quasi-invertible. In a unital ring this means that $1 - uv$ is invertible if and only if $1 - vu$ is invertible.

5.19. Let R be an arbitrary ring. Show that

$$\mathrm{rad}\left(\begin{bmatrix} R & R \\ 0 & R \end{bmatrix}\right) = \begin{bmatrix} \mathrm{rad}(R) & R \\ 0 & \mathrm{rad}(R) \end{bmatrix}.$$

Remark: Of course, this can be extended to upper triangular matrices over R of any size. It is also appropriate to mention the formula $\mathrm{rad}(M_n(R)) = M_n(\mathrm{rad}(R))$, which could as well be included among exercises. Its proof is not difficult, yet somewhat more involved than the one concerning the triangular case.

5.20. Show that a nonzero ring R is semiprimitive if and only if R has a faithful semisimple module.

5.21. Let R be a semiprimitive ring. True or False:

(a) If for every primitive ideal P of R the ring R/P has no nonzero nilpotent elements, then R has no nonzero nilpotent elements.

(b) If for every primitive ideal P of R the ring R/P has no nontrivial idempotents, then R has no nontrivial idempotents.

5.22. Let R be a semiprimitive ring. Show that the following statements are equivalent for $a \in R$:

(i) For every $x \in R$ there exists $m(x) \in \mathbb{N}$ such that $[a, x]^{m(x)} = 0$.

(ii) For every $x \in R$ there exists $n(x) \in \mathbb{N}$ such that $(ax)^{n(x)} = (xa)^{n(x)}$.

(iii) $a \in Z(R)$.

5.23. Show that an algebraic division algebra over a finite field is commutative.

Hint: One way of proving this result, which is also due to Jacobson, is to use Jacobson's Theorem 5.78, together with standard facts about fields and their extensions.

5.24. Show that a ring R is commutative if every commutator in R is an idempotent.

Remark: This is a very special case of Herstein's generalization of Theorem 5.78 which says a ring R is commutative if and only if for every pair $x, y \in R$ there exists an integer $n(x, y) \geq 2$ such that $[x, y]^{n(x,y)} = [x, y]$.

Chapter 6
Noncommutative Polynomials

Polynomials in several variables are a fundamental topic of interest in commutative algebra. Their counterparts in noncommutative algebra are the noncommutative polynomials, i.e., polynomials in noncommuting indeterminates. Thus, if ξ, η are such indeterminates, then $\xi\eta$ and $\eta\xi$ are different noncommutative polynomials. Operations on such polynomials are defined in the obvious way. For example, if $f = \xi\eta + \eta$ and $g = \eta\xi - \eta$, then

$$f + g = \xi\eta + \eta\xi, \quad fg = \xi\eta^2\xi + \eta^2\xi - \xi\eta^2 - \eta^2, \quad gf = \eta\xi^2\eta - \eta^2.$$

The purpose of this chapter is to give a basic introduction to a couple of topics related to noncommutative polynomials. There will be plenty of examples and simple observations, more than theorems and proofs. The emphasis will be on polynomial identities. These are noncommutative polynomials which vanish when replacing indeterminates with arbitrary elements from a given algebra. However, most of the deeper results on polynomial identities will be obtained in the next chapter, when we will have more tools at our disposal. The main goal of this chapter is to introduce an appropriate framework.

6.1 Free Algebras

After the above informal introduction, let us define noncommutative polynomials precisely. Take a nonempty set X. It is convenient to consider it as an indexed set, so let us write $X = \{\xi_i \mid i \in I\}$. A finite sequence of elements from X will be denoted by $\xi_{i_1}\xi_{i_2}\ldots\xi_{i_m}$ and called a **word**. The empty sequence is not excluded; we denote it by 1 and call it the **empty word**. Defining multiplication by juxtaposition, i.e.,

$$(\xi_{i_1}\xi_{i_2}\ldots\xi_{i_m}) \cdot (\xi_{j_1}\xi_{j_2}\ldots\xi_{j_n}) := \xi_{i_1}\xi_{i_2}\ldots\xi_{i_m}\xi_{j_1}\xi_{j_2}\ldots\xi_{j_n},$$

© Springer International Publishing Switzerland 2014
M. Brešar, *Introduction to Noncommutative Algebra*, Universitext,
DOI 10.1007/978-3-319-08693-4_6

the set of all words becomes a monoid (with unity 1), which we denote by X^*. It is called the **free monoid** on X. Viewing $\xi_i \in X$ as an element of X^*, we can write ξ_i^2 for $\xi_i\xi_i$, ξ_i^3 for $\xi_i\xi_i\xi_i$, etc. Accordingly, every $w \neq 1$ in X^* can be written as $w = \xi_{i_1}^{k_1}\xi_{i_2}^{k_2}\dots\xi_{i_r}^{k_r}$, where $i_j \in I$ and $k_j \in \mathbb{N}$. Of course, i_j can be equal to i_k if $j \neq k$, but we can achieve that $i_j \neq i_{j+1}$.

Recall that given a monoid and a field one can form a monoid algebra.

Definition 6.1 Let F be a field and X a set. The **free algebra** on X over F is the monoid algebra of X^* over F.

In short, this is the algebra whose basis is the set of all words formed from X (including 1) and whose multiplication is determined by juxtaposition of words.

The free algebra on X over F will be denoted by $F\langle X\rangle$, or, if X is a finite (resp. countably infinite) set, also by $F\langle\xi_1,\dots,\xi_n\rangle$ (resp. $F\langle\xi_1,\xi_2,\dots\rangle$). The elements in X are called **indeterminates**. If X has just two elements, then we will avoid indices and write $F\langle\xi,\eta\rangle$. We call elements in free algebras **noncommutative polynomials**. However, we will usually omit "noncommutative" and call them simply **polynomials**. The classical polynomials in commuting indeterminates will also occasionally appear in the sequel, but there is little chance for confusion. Already the notation will be different. We will use ω for denoting commuting indeterminates, and ξ or η for noncommuting indeterminates.

The notions such as a **coefficient**, the **constant term**, and a **constant polynomial** are defined for noncommutative polynomials in the same way as for commutative polynomials.

The role of the field F in the definition of $F\langle X\rangle$ is purely formal. One can replace it with an arbitrary ring R and form the ring $R\langle X\rangle$ of (noncommutative) polynomials with coefficients in R. The operations in $R\langle X\rangle$ are defined just as in $F\langle X\rangle$. In particular, elements from R commute with indeterminates from X, but not necessarily between themselves. Studying $R\langle X\rangle$ at a basic level is not much harder than studying $F\langle X\rangle$ with F a field. Nevertheless, we restrict ourselves to the latter, at least in this chapter. In the next one we will actually encounter the case where $R = \mathbb{Z}$. One can also describe $\mathbb{Z}\langle X\rangle$ as the subring of $\mathbb{Q}\langle X\rangle$ consisting of polynomials with integer coefficients.

Let A be another unital F-algebra. A *unital* algebra homomorphism (i.e., an algebra homomorphism that sends 1 into 1) from $F\langle X\rangle$ into A is uniquely determined by the action on X. This is clear since $F\langle X\rangle$ is generated by the set $X \cup \{1\}$. What we wish to point out is a kind of converse: An arbitrary function f from X into A can be extended to a unital algebra homomorphism $\bar{f}$ from $F\langle X\rangle$ into A. Indeed, there is only one way to define this extension, and it is well-defined and is a homomorphism. The following diagram is thus commutative.

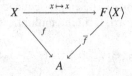

This may be just a simple observation, but is of fundamental importance. We will refer to it as to the **universal property** of $F\langle X\rangle$. (Readers familiar with the categorical language will recognize that this property shows that $F\langle X\rangle$ is a free object in the category of unital F-algebras.)

Every polynomial $f \in F\langle X\rangle$ involves only a finite number of indeterminates. If these are $\xi_{i_1}, \ldots, \xi_{i_n}$, then we write $f = f(\xi_{i_1}, \ldots, \xi_{i_n})$. Usually we will simply write $f = f(\xi_1, \ldots, \xi_n)$. Given such a polynomial and elements $x_1, \ldots, x_n \in A$, we define the **evaluation** $f(x_1, \ldots, x_n)$ in the obvious way. Say, if $f = \xi_1^2\xi_2\xi_1 - \xi_2\xi_3 + 1$, then $f(x_1, x_2, x_3) = x_1^2 x_2 x_1 - x_2 x_3 + 1 \in A$, i.e., one just substitutes x_i for ξ_i. In more formal terms, the universal property shows that the function $\xi_i \mapsto x_i$ gives rise to the unital homomorphism from $F\langle \xi_1, \ldots, \xi_n\rangle$ into A which maps f into $f(x_1, \ldots, x_n)$. If A is not unital, then $f(x_1, \ldots, x_n)$ still makes sense provided that f has zero constant term.

A polynomial is said to be a **monomial** if it is a nonzero scalar multiple of a word. Every nonzero polynomial $f \in F\langle X\rangle$ is a sum of monomials. More precisely, f can be uniquely written as $f = \lambda_1 w_1 + \cdots + \lambda_m w_m$ where $w_1, \ldots, w_m$ are pairwise different words and the coefficients $\lambda_1, \ldots, \lambda_m$ are nonzero scalars. Then f is the sum of the monomials $\lambda_i w_i$.

We define the **length** of a nonempty word $w = \xi_{i_1}\xi_{i_2}\ldots\xi_{i_m}$ as $\ell(w) := m$, and the length of the empty word as $\ell(1) := 0$. The **degree** of a nonzero polynomial $f = \lambda_1 w_1 + \cdots + \lambda_m w_m$ ($\lambda_i \neq 0$) is defined as

$$\deg(f) := \max\{\ell(w_1), \ldots, \ell(w_m)\}.$$

If $\ell(w_1) = \cdots = \ell(w_m)$, i.e., if all monomials of f have the same degree, then f is said to be a **homogeneous polynomial**. For instance, $\xi^2\eta - \xi\eta\xi + \eta^3$ is homogeneous of degree 3. We will be often concerned with homogeneous polynomials of the following type: $f = f(\xi_1, \ldots, \xi_n)$ is called a **multilinear polynomial** if each indeterminate ξ_i appears in every monomial of f exactly once. That is, f is of the form

$$f = \sum_{\sigma \in S_n} \lambda_\sigma \xi_{\sigma(1)} \ldots \xi_{\sigma(n)},$$

where $\lambda_\sigma \in F$ and S_n is the symmetric group on $\{1, \ldots, n\}$. Observe that for every F-algebra A, the map $(x_1, \ldots, x_n) \mapsto f(x_1, \ldots, x_n)$, $x_i \in A$, is multilinear.

Take nonzero polynomials $f = \lambda_1 w_1 + \cdots + \lambda_m w_m$ and $g = \mu_1 z_1 + \cdots + \mu_n z_n$, and assume, for notational simplicity, that $\deg(f) = \ell(w_1)$ and $\deg(g) = \ell(z_1)$. Note that $w_1 z_1 = w_k z_l$ occurs only when $k = l = 1$. Hence $\deg(fg) = \ell(w_1 z_1) = \ell(w_1) + \ell(z_1)$. Thus,

$$\deg(fg) = \deg(f) + \deg(g)$$

holds for all nonzero $f, g \in F\langle X\rangle$ (if one defines $\deg(0) := -\infty$, then "nonzero" can be omitted). An immediate consequence is that $F\langle X\rangle$ *is a domain*.

Remark 6.2 A much deeper result is that $F\langle X \rangle$ *is a primitive algebra* unless X is a singleton (note that $F\langle \xi \rangle$ is just the usual polynomial algebra $F[\xi]$ which is not primitive by Lemma 5.7). We shall not need this result, so we omit the proof. Let us just give its rough outline for the simplest case where $X = \{\xi, \eta\}$. Take a vector space V over F with basis $\{e_n \mid n \in \mathbb{N}\}$. Define $g, h \in \mathrm{End}_F(V)$ by $g(e_1) = 0$ and $g(e_n) = e_{n-1}$ for $n \neq 1$, and $h(e_n) = e_{n^2+1}$ for every n. Let A be the subalgebra of $\mathrm{End}_F(V)$ generated by g and h. It is not hard to prove that A is a primitive algebra, in fact a dense algebra of linear operators of V. A slightly harder, but still elementary exercise is to prove that $F\langle \xi, \eta \rangle$ is isomorphic to A. The point is to show that the homomorphism $f(\xi, \eta) \mapsto f(g, h)$, which exists by the universal property, is injective. Details can be found, for example, in [Lam01, pp. 184–185].

6.2 Algebras Defined by Generators and Relations

Let A be a unital algebra over a field F. Take a subset X of A that generates A. In principle, X can be any set of generators, but in practice we are more interested in "smaller" sets. Forget for the moment that X is a subset of an algebra and consider it merely as a set. Then we can form the free algebra $F\langle X \rangle$. The function $\xi \mapsto \xi$ from the set X into the algebra A can be, as pointed out in the preceding section, uniquely extended to a unital algebra homomorphism φ from $F\langle X \rangle$ into A. As its range contains X, φ is surjective. We have thereby shown that *every unital algebra is a homomorphic image of a free algebra*. Setting $I = \ker \varphi$ we can rephrase this as follows: For every unital F-algebra A there exists a free algebra $F\langle X \rangle$ and its ideal I such that

$$A \cong F\langle X \rangle / I.$$

These simple observations indicate that free algebras play a special role in non-commutative algebra. They are hidden behind every algebra. When searching for counterexamples, one often starts by examining free algebras.

Let us now change the viewpoint. We begin by taking a nonempty set X and a nonempty subset R of $F\langle X \rangle$. Let us form the factor algebra $F\langle X \rangle / (R)$ where (R) is the ideal of $F\langle X \rangle$ generated by R. As noticed above, every unital algebra is isomorphic to an algebra of this kind. This construction thus gives us endless opportunities to create interesting examples. Also, it makes it possible for us to get a clearer picture of some important algebras that we often come across. Before proceeding to examples, we introduce some notation and terminology. We set $X = \{\xi_i \mid i \in I\}$ and $R = \{f_j = f_j \left(\xi_{i_1}, \ldots, \xi_{i_{n(j)}} \right) \mid j \in J\}$. Denote the coset $\xi_i + (R)$ by x_i. Note that $f_j \left(x_{i_1}, \ldots, x_{i_{n(j)}} \right) = 0$ for every $j \in J$. Let us switch to a more suggestive notation and write $F\langle X \rangle / (R)$ as

$$F\langle x_i, \ i \in I \mid f_j \left(x_{i_1}, \ldots, x_{i_{n(j)}} \right) = 0, j \in J \rangle.$$

We say that this algebra is defined by the **generators** x_i and **relations** f_j. Indeed the elements $x_i, i \in I$, together with 1 generate the algebra, and, unlike the generators ξ_i of the free algebra $F\langle X \rangle$, they are not independent but related through the polynomials f_j. These may not be the only polynomials that relate the x_i's. However, if $f = f(\xi_{i_1}, \ldots, \xi_{i_n}) \in F\langle X \rangle$ is such that $f(x_{i_1}, \ldots, x_{i_n}) = 0$, i.e.,

$$f(\xi_{i_1}, \ldots, \xi_{i_n}) + (R) = f(\xi_{i_1} + (R), \ldots, \xi_{i_n} + (R)) = 0,$$

then f lies in the ideal of $F\langle X \rangle$ generated by the f_j's.

Example 6.3 Let $A := F\langle x, y \,|\, x^2 = 0, \ y^2 = 0, \ xy + yx = 1 \rangle$. A concrete example of a pair of elements satisfying these relations are the matrix units E_{12} and E_{21} in $M_2(F)$. Let us show that A is actually isomorphic to $M_2(F)$. We begin by noticing that $xyx = x$ and $yxy = y$. Hence it follows that every element in A is a linear combination of x, y, xy, and yx. Suppose $\lambda_1 x + \lambda_2 y + \lambda_3 xy + \lambda_4 yx = 0$ for some $\lambda_i \in F$. Then the polynomial $f(\xi, \eta) = \lambda_1 \xi + \lambda_2 \eta + \lambda_3 \xi \eta + \lambda_4 \eta \xi$ lies in the ideal of $F\langle \xi, \eta \rangle$ generated by $R = \{\xi^2, \eta^2, \xi \eta + \eta \xi - 1\}$. Accordingly, for every F-algebra B and every pair of elements $a, b \in B$, the evaluation $f(a, b)$ lies in the ideal of B generated by the set $\{a^2, b^2, ab + ba - 1\}$. Choosing $B = M_2(F)$ and $a = E_{12}, b = E_{21}$, we see that each $\lambda_i = 0$. Consequently, $\{x, y, xy, yx\}$ is a basis of A. Note that its multiplication table is the same as that of $\{E_{12}, E_{21}, E_{11}, E_{22}\}$ in $M_2(F)$. Therefore $A \cong M_2(F)$.

Example 6.4 Assume that $\mathrm{char}(F) = 0$. Let us show that the first Weyl algebra $\mathscr{A}_1$ over F, introduced in Example 1.13, is isomorphic to $A := F\langle x, y \,|\, [x, y] = 1 \rangle$. By the universal property of $F\langle \xi, \eta \rangle$ we can define the unital homomorphism $f(\xi, \eta) \mapsto f(D, L)$ from $F\langle \xi, \eta \rangle$ onto $\mathscr{A}_1$. Since $R := \{[\xi, \eta] - 1\}$ is contained in its kernel, this homomorphism gives rise to the homomorphism $f(x, y) \mapsto f(D, L)$ from $A = F\langle \xi, \eta \rangle / (R)$ onto $\mathscr{A}_1$. We have to show that this is an isomorphism. To this end, we refer to the proof of the assertion (a) from Example 1.13. It is evident that it depends only on the formula $[D, L] = I$. Therefore the same conclusion holds for every unital algebra generated by a pair of elements whose commutator equals 1. Accordingly, every element in A can be written as a linear combination of the elements $y^m x^n$, $m, n \geq 0$. Using (b) from Example 1.13 we now see that our homomorphism is injective, and hence an isomorphism.

Now take an arbitrary $n \in \mathbb{N}$. The nth **Weyl algebra**, $\mathscr{A}_n$, is defined as the subalgebra of $\mathrm{End}_F(F[\omega_1, \ldots, \omega_n])$, the algebra of all linear operators of $F[\omega_1, \ldots, \omega_n]$, generated by the operators $D_i, L_i, i = 1, \ldots, n$, given by

$$D_i\big(f(\omega_1, \ldots, \omega_n)\big) = \frac{\partial f}{\partial \omega_i}(\omega_1, \ldots, \omega_n), \quad L_i\big(f(\omega_1, \ldots, \omega_n)\big) = \omega_i f(\omega_1, \ldots, \omega_n).$$

One can show that $\mathscr{A}_n$ is isomorphic to

$$F\langle x_i, y_i, \ i = 1, \ldots, n \,|\, [x_i, y_j] = \delta_{ij}, \ [x_i, x_j] = 0, \ [y_i, y_j] = 0, \ i, j = 1, \ldots, n \rangle,$$

and that $\mathscr{A}_n$ is, just as $\mathscr{A}_1$ (cf. Examples 1.13, 2.28, and 3.68), a simple noetherian domain. Proving all this for an arbitrary n is not much harder than for $n = 1$.

To provoke the reader's curiosity we mention a problem by J. Dixmier from 1968.

Problem 6.5 Is every nonzero algebra endomorphism of $\mathscr{A}_n$ surjective?

Recently it has turned out that this problem, to which one usually refers to as **Dixmier's conjecture**, is equivalent to the famous **Jacobian conjecture** from algebraic geometry. We omit stating this conjecture here; the interested reader will easily find information about it at many places.

Example 6.6 The algebra $A := F\langle x_i, i \in I \mid [x_i, x_j] = 0, i, j \in I\rangle$ is generated by the elements x_i which commute with each other, and these commuting relations are the only defining relations. It should come as no surprise that A is the same thing as $F[\omega_i, i \in I]$, the algebra of polynomials in commuting indeterminates ω_i. To give a formal proof, we use the universal property to define the unital homomorphism from $F\langle X \rangle, X = \{\xi_i \mid i \in I\}$, onto $F[\omega_i, i \in I]$, determined by $\xi_i \mapsto \omega_i$. Its kernel contains the set $R := \{\xi_i\xi_j - \xi_j\xi_i \mid i, j \in I\}$, and every element in $A = F\langle X \rangle/(R)$ can be written as $f + (R)$ where f is such that in each of its monomials the indeterminates ξ_i appear in the same fixed order. This readily implies that $A \cong F[\omega_i, i \in I]$.

Example 6.7 The **Grassmann algebra** (or the **exterior algebra**) over a field F with $\text{char}(F) \neq 2$ is defined as

$$G := F\langle x_i, i \in \mathbb{N} \mid x_i^2 = 0, x_ix_j + x_jx_i = 0, i, j \in \mathbb{N}\rangle.$$

Hence we have, for example, $x_1x_2x_1 = -x_1^2x_2 = 0$; in general, $x_iGx_i = 0$. Note also that every product of different x_i's can be written as plus or minus

$$x_{i_1}x_{i_2}\ldots x_{i_n}, \quad i_1 < i_2 < \cdots < i_n. \tag{6.1}$$

Accordingly, these elements, together with 1, linearly span G. Let us show that each of them is nonzero. It suffices to show that $x_1x_2\ldots x_n \neq 0$. If this was not true, then $\xi_1\xi_2\ldots\xi_n$ would be contained in the ideal of $F\langle\xi_1, \xi_2, \ldots\rangle$ generated by $R = \{\xi_i^2, \xi_i\xi_j + \xi_j\xi_i \mid i, j \in \mathbb{N}\}$. Then we could write $\xi_1\xi_2\ldots\xi_n$ as a sum of terms of the form $m\xi_i^2m'$ and $p(\xi_i\xi_j + \xi_j\xi_i)p'$, where m, m', p, p' are monomials. The linear span of all multilinear monomials, i.e., the space of all multilinear polynomials, has trivial intersection with the linear span of all monomials that are not multilinear. Hence it follows that $\xi_1\xi_2\ldots\xi_n$ is a linear combination of polynomials

$$\xi_{\sigma(1)} \cdots \xi_{\sigma(j-1)}\Big(\xi_{\sigma(j)}\xi_{\sigma(j+1)} + \xi_{\sigma(j+1)}\xi_{\sigma(j)}\Big)\xi_{\sigma(j+2)} \cdots \xi_{\sigma(n)}$$

where $\sigma \in S_n$. Such a linear combination is a multilinear polynomial with the property that the sum of its coefficients corresponding to even permutations coincides with the sum of its coefficients corresponding to odd permutations. Since $\xi_1\xi_2\ldots\xi_n$

does not share this property, we have arrived at a contradiction. Thus, the elements from (6.1) are indeed nonzero.

Suppose $\sum_{i=1}^{n} \lambda_i b_i = 0$ where $\lambda_i \in F \backslash \{0\}$ and each b_i is either 1 or is as in (6.1). We may assume that there exists $k \in \mathbb{N}$ such that x_k appears in b_n, but not in b_1. Hence $\sum_{i=1}^{n-1} \lambda_i b_i x_k = 0$ and $b_1 x_k \neq 0$. Continuing this procedure we arrive at a contradiction that one of the λ_i's is zero. Thus, the elements of the form (6.1), together with 1, form a basis of G.

It is obvious that $x_1 x_2$ commutes with each x_i, and hence it lies in $Z(G)$, the center of G. The same reasoning shows that every element from (6.1) with n even lies in $Z(G)$. Moreover, using the assumption that char$(F) \neq 2$ it is easy to see that $Z(G)$ is actually the linear span of all such elements and 1. We will write G_0 for $Z(G)$, and similarly, we will write G_1 for the linear span of all elements of the form (6.1) with n odd. Note that $G = G_0 \oplus G_1$, the vector space direct sum.

One often considers the Grassmann algebra in n generators. We can describe it as the subalgebra of G generated by $1, x_1, \ldots, x_n$. By counting the basis elements we see that its dimension is 2^n.

The Grassmann algebra has a big center and plenty of nilpotent ideals, in contrast to algebras that we prefer in structure theory. However, G has turned out to be an extremely useful tool in various mathematical areas. A small evidence of its applicability will be given in the proof of the Amitsur-Levitzki Theorem in Sect. 6.9.

6.3   Alternating Polynomials

In this section we consider special polynomials that play an important role in the theory of polynomial identities, the theme of the subsequent sections. For simplicity we will confine ourselves to multilinear polynomials.

Let us begin with an example. Consider the polynomial

$$h(\xi_1, \xi_2, \xi_3, \eta) = \xi_1 \eta \xi_2 \xi_3 - \xi_1 \eta \xi_3 \xi_2 + \xi_2 \eta \xi_3 \xi_1 - \xi_2 \eta \xi_1 \xi_3 + \xi_3 \eta \xi_1 \xi_2 - \xi_3 \eta \xi_2 \xi_1.$$

Note that by substituting ξ_1 for ξ_2 we get 0, i.e., $h(\xi_1, \xi_1, \xi_3, \eta) = 0$. Similarly, $h(\xi_1, \xi_2, \xi_1, \eta) = 0$ and $h(\xi_1, \xi_2, \xi_2, \eta) = 0$.

Definition 6.8 A multilinear polynomial $f = f(\xi_1, \ldots, \xi_n, \eta_1, \ldots, \eta_r) \in F\langle X \rangle$ is said to be **alternating** in $\xi_1, \ldots, \xi_n$ if f becomes zero whenever one substitutes ξ_j for ξ_i, $1 \leq i < j \leq n$.

Replacing ξ_1 by $\xi_1 + \xi_2$ in $f(\xi_1, \xi_1, \xi_3, \ldots, \xi_n, \eta_1, \ldots, \eta_r) = 0$ we get

$$f(\xi_1, \xi_2, \xi_3, \ldots, \xi_n, \eta_1, \ldots, \eta_r) = -f(\xi_2, \xi_1, \xi_3, \ldots, \xi_n, \eta_1, \ldots, \eta_r).$$

Similarly, f changes sign if we switch any pair of indeterminates ξ_i and ξ_j. The name "alternating" comes from this condition, which is obviously equivalent to the definition provided that char$(F) \neq 2$.

Our main reason for considering alternating polynomials is their connection to the notion of linear dependence.

Lemma 6.9 *Let A be an F-algebra and let $a_1, \ldots, a_n \in A$ be linearly dependent elements. If a multilinear polynomial $f = f(\xi_1, \ldots, \xi_n, \eta_1, \ldots, \eta_r) \in F\langle X \rangle$ is alternating in $\xi_1, \ldots, \xi_n$, then $f(a_1, \ldots, a_n, x_1, \ldots, x_r) = 0$ for all $x_1, \ldots, x_r \in A$.*

Proof We may assume that, say, $a_n = \sum_{i=1}^{n-1} \lambda_i a_i$ for some $\lambda_i \in F$. Consequently,

$$f(a_1, \ldots, a_n, x_1, \ldots, x_r) = \sum_{i=1}^{n-1} \lambda_i f(a_1, \ldots, a_{n-1}, a_i, x_1, \ldots, x_r) = 0,$$

since $f(\xi_1, \ldots, \xi_i, \ldots, \xi_{n-1}, \xi_i, \eta_1, \ldots, \eta_r) = 0$ for $i = 1, \ldots, n-1$.   □

Two families of alternating polynomials are of special importance. The first one consists of polynomials that are alternating in all indeterminates (i.e., there are no η_j's).

Definition 6.10 Let $n \geq 2$. The polynomial

$$s_n = s_n(\xi_1, \ldots, \xi_n) := \sum_{\sigma \in S_n} \mathrm{sgn}(\sigma) \xi_{\sigma(1)} \ldots \xi_{\sigma(n)}$$

is called the **standard polynomial** of degree n.

The reader's first thought may have been that this is similar to the definition of the determinant of a matrix. There are indeed several analogies.

Choose indices i, j and a permutation σ. Write $\xi_{\sigma(1)} \ldots \xi_{\sigma(n)}$ as $m\xi_i m' \xi_j m''$ where m, m', m'' are monomials in the remaining indeterminates. Then the term $m\xi_j m' \xi_i m''$ corresponds to a permutation whose sign is opposite to that of σ. This readily implies that s_n is alternating.

Note that $s_2 = \xi_1 \xi_2 - \xi_2 \xi_1$, and $s_3 = h(\xi_1, \xi_2, \xi_3, 1)$ where h is the above polynomial (substituting 1 for η simply means that one can "erase" η in the formula defining h).

The second family is similar, but it involves the η_j's.

Definition 6.11 Let $n \geq 2$. The polynomial

$$c_n = c_n(\xi_1, \ldots, \xi_n, \eta_1, \ldots, \eta_{n-1}) := \sum_{\sigma \in S_n} \mathrm{sgn}(\sigma) \xi_{\sigma(1)} \eta_1 \xi_{\sigma(2)} \eta_2 \cdots \eta_{n-1} \xi_{\sigma(n)}$$

is called the **nth Capelli polynomial**.
(Warning: some authors use slightly different definitions.)

Thus, $c_2 = \xi_1 \eta_1 \xi_2 - \xi_2 \eta_1 \xi_1$, and one gets c_3 by writing η_1 for η and inserting η_2 between the last two factors of each term of h. Note that

$$s_n(\xi_1, \ldots, \xi_n) = c_n(\xi_1, \ldots, \xi_n, 1, \ldots, 1).$$

It is easy to see that c_n is alternating in $\xi_1, \ldots, \xi_n$. Let us also mention the easily derived formula

$$c_n = \sum_{i=1}^{n} (-1)^{i-1} \xi_i \eta_1 c_{n-1}(\xi_1, \ldots, \xi_{i-1}, \xi_{i+1}, \ldots, \xi_n, \eta_2, \ldots, \eta_{n-1}), \qquad (6.2)$$

which holds for every $n \geq 3$. Similarly, for every $n \geq 3$ we have

$$s_n = \sum_{i=1}^{n} (-1)^{i-1} \xi_i s_{n-1}(\xi_1, \ldots, \xi_{i-1}, \xi_{i+1}, \ldots, \xi_n). \qquad (6.3)$$

6.4 Polynomial Identities: Definition and Examples

Much of the rest of the book is devoted to the following notion.

Definition 6.12 A polynomial $f = f(\xi_1, \ldots, \xi_n) \in F\langle \xi_1, \xi_2, \ldots \rangle$ is said to be a **polynomial identity** (or simply an **identity**) of an F-algebra A if $f(x_1, \ldots, x_n) = 0$ for all $x_1, \ldots, x_n \in A$. In this case we also say that A **satisfies** f.

The reader should not confuse this with algebras defined by generators and relations, where polynomials vanish only on some of the generators. The condition that f is an identity of A is quite restrictive, provided, of course, that f is a nonzero polynomial.

Definition 6.13 If a nonzero polynomial is a polynomial identity of A, then A is called a PI-**algebra**.

There are two basic examples of PI-algebras.

Example 6.14 Saying that A is commutative is the same as saying that A satisfies the identity $[\xi_1, \xi_2]$ $(= \xi_1\xi_2 - \xi_2\xi_1)$. Every commutative algebra is thus a PI-algebra.

This suggests one way of looking at the notion of a polynomial identity, i.e., as a generalization of commutativity. The second basic example is of a different nature.

Example 6.15 Every finite dimensional algebra is a PI-algebra. If $[A : F] = d$, then every multilinear polynomial that is alternating in $d + 1$ indeterminates is an identity of A by Lemma 6.9. In particular, A satisfies s_{d+1}.

Thus, $M_n(F)$ is a PI-algebra. In fact, a considerable part of the theory of polynomial identities is centered around matrices. Roughly speaking, the theory has two branches, *combinatorial* and *structural*. The goal of the first one is understanding of the set of all identities of a given algebra A, with a special emphasis on the case where $A = M_n(F)$. Concerning the second one, let us only mention at this point that a prime PI-algebra can be embedded, in a particularly nice way, into a matrix algebra over a field extension of the base field. This indicative result is a byproduct of the structure theory which will be exposed in the next chapter (see Corollary 7.59 and Remark 7.60). The combinatorial aspect will be briefly touched upon in the rest of this chapter.

There are other examples of PI-algebras, quite different from commutative and finite dimensional ones.

Example 6.16 The commutator $[x, y]$ of any two elements x, y in the Grassmann algebra G lies in $Z(G) = G_0$. It is enough to show this for basis elements $x = x_{i_1} \ldots x_{i_n}$ and $y = x_{j_1} \ldots x_{j_m}$. Note that $[x, y]$ can be nonzero only if both n and m are odd, and in this case it indeed lies in G_0. Thus, G satisfies the identity $[[\xi_1, \xi_2], \xi_3]$.

On the other hand, if char$(F) = 0$, then none of the standard polynomials is an identity of G. To show this, we first observe that

$$x_{\sigma(1)} \ldots x_{\sigma(r)} = \text{sgn}(\sigma)x_1 \ldots x_r \tag{6.4}$$

holds for every permutation σ on $\{1, \ldots, r\}$. By writing σ as a product of transpositions we see that it is enough to prove (6.4) for the case where σ itself is a transposition, and in this case the formula is pretty obvious. Consequently, $s_r(x_1, \ldots, x_r) = r!x_1 \ldots x_r \neq 0$ for every $r \geq 2$. From Example 6.15 we therefore infer that G cannot be embedded into any algebra that is finite dimensional over some field extension of F; in fact, it cannot even be embedded into $M_n(C)$ for any commutative algebra C (see Theorem 6.39 below).

Example 6.17 If A is a nilpotent algebra, then A satisfies the identity $\xi_1 \ldots \xi_n$ for some $n \in \mathbb{N}$. A concrete example is the algebra of all *strictly* upper triangular matrices over a commutative algebra C. Now consider $T_n(C)$, the algebra of all upper triangular $n \times n$ matrices over C. Note that the commutator of any two matrices from $T_n(C)$ is a strictly upper triangular matrix. Therefore $T_n(C)$ satisfies $[\xi_1, \eta_1] \ldots [\xi_n, \eta_n]$.

Example 6.18 Suppose ξ^2 is an identity of A. This does not necessarily mean that $A^2 = 0$ (for example, consider the 3-dimensional subalgebra $Fx_1 + Fx_2 + Fx_1x_2$ of the Grassmann algebra G). However, $A^3 = 0$, provided that char$(F) \neq 2$. The proof is easy. We have $x^2 = 0$ for every $x \in A$. Replacing x by $x + y$, and using $x^2 = y^2 = 0$, we get $xy + yx = 0$ for all $x, y \in A$ (i.e., $\xi\eta + \eta\xi$ is an identity of A). Consequently, for all $x, y, z \in A$ we have

$$(xy)z = -z(xy) = -(zx)y = y(zx) = (yz)x = -x(yz),$$

yielding $xyz = 0$.

Much more can be said. If char$(F) = 0$ and A satisfies the identity ξ^n, then A is a nilpotent algebra, i.e., A satisfies $\xi_1\xi_2 \ldots \xi_d$ for some d, depending on n. This result is called the **Nagata-Higman Theorem**. The proof uses elementary tools, but we omit it (it can be found, for example, in [ZSSS82]). Finding the exact lower bound for d is an open problem.

Example 6.19 Let $n \geq 2$. Theorem 5.78 implies that an algebra satisfying the identity $\xi^n - \xi$ is commutative. Concrete examples are Boolean rings and finite fields. Of course, Theorem 5.78 says more than that. However, the condition $x^{n(x)} = x$ cannot be interpreted in terms of polynomial identities.

How to obtain new PI-algebras from the old ones? In the next two examples we give some answers to this question. The first example, as trivial as it is, is of fundamental importance.

Example 6.20 If f is an identity of an algebra A, then f is also an identity of every subalgebra of A as well as of every homomorphic image of A (equivalently, of A/I for every ideal I of A). In particular, a subalgebra or a homomorphic image of a PI-algebra is again a PI-algebra.

Example 6.21 Let $\{A_i \mid i \in I\}$ be a family of algebras. If each A_i satisfies the same identity f, then so does the direct product $\Pi_{i \in I} A_i$. On the other hand, if each A_i is a PI-algebra, then there is no reason to believe that $\Pi_{i \in I} A_i$ is a PI-algebra too. If, however, the set I is finite, say $I = \{1, \dots, n\}$, then this is true. Namely, if f_i is an identity of A_i, then $f_1 \dots f_n$ is an identity of $A_1 \times \cdots \times A_n$. Not only the direct product, even the tensor product of two PI-algebras is again a PI-algebra. But this is anything but obvious. It was proved by A. Regev in the early 1970s.

All these examples may give the wrong impression that most algebras are PI-algebras. Infinite dimensional algebras with small centers, e.g., infinite dimensional central algebras, are actually only rarely PI-algebras. This will become clear as the theory unfolds. Let us mention, at this point, just an obvious example of an algebra without nonzero identities.

Example 6.22 The free algebra $F\langle \xi_1, \xi_2, \dots \rangle$ is not a PI-algebra.

One can define polynomial identities for rings, not only for algebras. We will in fact deal with PI-rings in the next chapter. In this one, however, we consider only algebras in order to make the exposition simpler.

6.5 Linearization

The purpose of this section is to get familiar with the **process of linearization**. The concept of this process is applicable to various situations throughout mathematics. Our goal, however, is to show that it can be used to recast an arbitrary polynomial identity into a multilinear one.

In Example 6.18 we have already met a simple example of a linearization: If A satisfies ξ^2, then it also satisfies the multilinear identity $\xi\eta + \eta\xi$. The proof is very easy, one just replaces an arbitrary element by the sum of two arbitrary elements. This simple idea, which is the essence of "linearization", can be pushed further.

Example 6.23 Consider a slightly more complicated situation where A satisfies $f(\xi) = \xi^3$. All relevant calculations can be made in a free algebra, so we can ignore A for a while. Let us define two new polynomials $g = g(\xi_1, \xi_2)$ and $h = h(\xi_1, \xi_2, \xi_3)$ as follows:

$$g := f(\xi_1 + \xi_2) - f(\xi_1) - f(\xi_2),$$

$$h := g(\xi_1, \xi_2 + \xi_3) - g(\xi_1, \xi_2) - g(\xi_1, \xi_3).$$

Of course, h can be explicitly expressed by f, i.e.,

$$h = f(\xi_1 + \xi_2 + \xi_3) - f(\xi_1 + \xi_2) - f(\xi_1 + \xi_3) - f(\xi_2 + \xi_3) + f(\xi_1) + f(\xi_2) + f(\xi_3).$$

However, to compute h it is easier to first compute g,

$$g = \xi_1\xi_2^2 + \xi_2\xi_1\xi_2 + \xi_2^2\xi_1 + \xi_1^2\xi_2 + \xi_1\xi_2\xi_1 + \xi_2\xi_1^2,$$

and hence derive

$$h = \sum_{\sigma \in S_3} \xi_{\sigma(1)}\xi_{\sigma(2)}\xi_{\sigma(3)}.$$

Since A satisfies f, it also satisfies h. We have thus found a multilinear polynomial identity of A, but at the cost of increasing the number of indeterminates.

To consider arbitrary polynomials, we first introduce an auxiliary notion. The **degree of a polynomial** f **in** ξ_i is the maximal number of occurrences of ξ_i in the monomials of f. For example, $f = \xi_1^2\xi_2\xi_3\xi_1 + \xi_2\xi_3\xi_2$ has degree 3 in ξ_1, degree 2 in ξ_2, and degree 1 in ξ_3.

Theorem 6.24 *If an algebra A satisfies a nonzero polynomial identity, then A also satisfies a nonzero multilinear polynomial identity of the same or lower degree.*

Proof Let $f = f(\xi_1, \ldots, \xi_n)$ be a nonzero identity of A. Denote by d_i the degree of f in ξ_i. The proof is by induction on $d := \max\{d_1, \ldots, d_n\} > 0$.

If $d = 1$, then each ξ_i appears in every monomial of f at most once—"at most" and not "exactly", so f is not necessarily multilinear. But this can be easily remedied. If, say, $\lambda\xi_1 \ldots \xi_m$, where λ is a nonzero scalar, is a monomial of f of minimal degree, then we just substitute 0 for the remaining indeterminates, and hence obtain a nonzero multilinear identity $f(\xi_1, \ldots, \xi_m, 0, \ldots, 0)$ of degree less or equal to $\deg(f)$.

Let $d > 1$. Without loss of generality we may assume that there exists $k \le n$ such that $d_k = \cdots = d_n = d$ and $d_i < d$ for $i < k$. Define a new polynomial which involves an additional indeterminate, $g = g(\xi_1, \ldots, \xi_n, \xi_{n+1})$, by

$$g := f(\xi_1, \ldots, \xi_{n-1}, \xi_n + \xi_{n+1}) - f(\xi_1, \ldots, \xi_{n-1}, \xi_n) - f(\xi_1, \ldots, \xi_{n-1}, \xi_{n+1}).$$

Note that g is also an identity of A. Let us write $f = \sum_i \lambda_i w_i$, where the w_i's are pairwise different words and the λ_i's are nonzero scalars. Then g can be written as $g = \sum_i \lambda_i g_i$, where g_i is obtained from w_i in the same way as g is obtained from f. If ξ_n does not occur in w_i, then $g_i = -w_i$. If ξ_n occurs only once in w_i, then $g_i = 0$. If ξ_n occurs at least twice in w_i, then g_i is the sum of all possible words obtained by replacing at least one, but not all, of the ξ_n's in w_i by ξ_{n+1}. Thus, if in any of these words one substitutes ξ_n for ξ_{n+1}, then one gets back the word w_i. Therefore the words appearing in g_i are different from the words appearing in $g_{i'}$ whenever $i' \neq i$. As $d > 1$, the indices i such that ξ_n occurs at least twice in w_i do exist. This shows that $g \neq 0$.

The following conclusions can be drawn from the preceding paragraph:

- g is a nonzero identity of A.
- $\deg(g) \leq \deg(f)$.
- For $j = 1, \ldots, n - 1$, the degree of g in ξ_j is less or equal to d_j.
- The degree of g in ξ_n and ξ_{n+1} is $d - 1$.

Now repeat this procedure, first with g in place of f and ξ_{n-1} in place of ξ_n, and then with other indeterminates down to ξ_k. At the end we arrive at a situation where a nonzero identity has degree at most $d - 1$ in every indeterminate. $\qquad\square$

This theorem is of crucial importance if one is interested in the structure of an algebra satisfying some nonzero polynomial identity. With regard to particular identities, it should be remarked that in the finite characteristic case some information can be lost when performing the linearization process. Take our first example where an F-algebra A satisfies ξ^2. Linearizing we get that it also satisfies $\xi\eta + \eta\xi$. If $\mathrm{char}(F) \neq 2$, then the two identities are equivalent, each one implies the other one. If $\mathrm{char}(F) = 2$, then the multilinear one only tells us that the algebra is commutative, while the original one shows much more, every element has square zero.

6.6 Stable Identities

Let A be an F-algebra and let $f \in F\langle X\rangle$ be an identity of A. Take an extension field K of F. Is f an identity of $A_K = K \otimes A$, the scalar extension of A to K? Since A_K has the same multiplication table as A, an affirmative answer seems plausible. However, the next simple example shows that this may not be true even for $A = F$.

Example 6.25 Let F be a finite field with n elements. Then $f = \xi^n - \xi$ is an identity of F (see Lemma 5.76). Since f cannot have more than n zeros in any field, f is not an identity of any extension field K of F that properly contains F.

The main goal of this section is to show that the above question does have an affirmative answer if F is infinite. Moreover, we may replace the role of the field K by an arbitrary commutative algebra C.

Definition 6.26 An identity f of an F-algebra A is said to be **stable** (for A) if f is an identity of $C \otimes A$ for every commutative F-algebra C.

Example 6.25 thus shows that $\xi^n - \xi$ is not a stable identity for $A = F$ if $|F| = n$. Let us turn to positive results.

Lemma 6.27 *Every multilinear identity is stable.*

Proof Let $f = f(\xi_1, \dots, \xi_n)$ be a multilinear polynomial, and consider, for arbitrary A and $C, f(\overline{x}_1, \dots, \overline{x}_n)$ with $\overline{x}_i \in C \otimes A$. Each $\overline{x}_i$ is a sum of simple tensors, and so, by multilinearity of f, $f(\overline{x}_1, \dots, \overline{x}_n)$ can be written as a sum of elements of the form $f(c_1 \otimes x_1, \dots, c_n \otimes x_n)$. Since C is commutative we have

$$f(c_1 \otimes x_1, \dots, c_n \otimes x_n) = c_1 \dots c_n \otimes f(x_1, \dots, x_n),$$

which is 0 if f is an identity of A. $\qquad\qquad\qquad\qquad\qquad\qquad\qquad \square$

The classical polynomials in commuting indeterminates play a similar role in commutative algebra as the noncommutative polynomials do in noncommutative algebra. The algebra of commutative polynomials $F[\Omega]$ satisfies the *universal property* in the context of commutative unital algebras, the *evaluation* of a commutative polynomial in elements from a commutative algebra can be defined in the obvious way, and every commutative unital algebra is a homomorphic image of $F[\Omega]$ for some set Ω. A detailed explanation of all this is essentially the same as for $F\langle X \rangle$ in Sects. 6.1 and 6.2. At any rate, these are just elementary observations—but very useful ones, as we shall see in the rest of the chapter.

Lemma 6.28 *Let $f = f(\xi_1, \dots, \xi_n)$ be an identity of an F-algebra A. If f is an identity of $F[\omega_1, \dots, \omega_s] \otimes A$ for every $s \in \mathbb{N}$, then f is stable for A.*

Proof Let C be a commutative unital F-algebra. Since every commutative algebra is a subalgebra of a commutative unital algebra (say, of its unitization), it suffices to show that f is an identity of $C \otimes A$. Take arbitrary elements $\overline{x}_i = \sum_{j=1}^{m_i} c_{ij} \otimes x_{ij}$, $i = 1, \dots, n$, in $C \otimes A$. Let us set

$$\tilde{x}_i := \sum_{j=1}^{m_i} \omega_{ij} \otimes x_{ij} \in F[\Omega] \otimes A,$$

where $\Omega = \{\omega_{ij} \mid i = 1, \dots, n, j = 1, \dots, m_i\}$. By assumption, $f(\tilde{x}_1, \dots, \tilde{x}_n) = 0$. Therefore, if φ is the (unital) homomorphism from $F[\Omega]$ into C sending ω_{ij} into c_{ij}, then $f(\overline{x}_1, \dots, \overline{x}_n) = (\varphi \otimes \mathrm{id}_A)\big(f(\tilde{x}_1, \dots, \tilde{x}_n)\big) = 0$. $\qquad \square$

Let us warn the reader that henceforth our argumentation will be somewhat less formal than in the proof of Lemma 6.28. We will avoid speaking about the universal property and homomorphisms from $F[\Omega]$, but simply substitute elements from commutative algebras for indeterminates in formulas involving polynomials.

Theorem 6.29 *Let F be an infinite field. Then every polynomial identity of an arbitrary F-algebra A is stable.*

Proof Let $f = f(\xi_1, \ldots, \xi_n)$ be an identity of A. Pick $s \in \mathbb{N}$ and take arbitrary elements $\hat{x}_i = \sum_j p_{ij} \otimes x_{ij}$ in $F[\omega_1, \ldots, \omega_s] \otimes A$, $i = 1, \ldots, n$. By Lemma 6.28 it is enough to show that $f(\hat{x}_1, \ldots, \hat{x}_n) = 0$. Suppose this is not true. Then we can write

$$f(\hat{x}_1, \ldots, \hat{x}_n) = \sum_k q_k \otimes a_k, \qquad (6.5)$$

where $q_1 \neq 0$ and the a_k's are linearly independent (cf. Remark 4.15). Since F is infinite, there exist $\lambda_1, \ldots, \lambda_s \in F$ such that $q_1(\lambda_1, \ldots, \lambda_s) \neq 0$ (the proof is a straightforward induction on s; for $s = 1$ just use the fact that the number of zeros of a polynomial cannot exceed its degree). Let us set $\alpha_{ij} := p_{ij}(\lambda_1, \ldots, \lambda_s)$ and $\beta_k := q_k(\lambda_1, \ldots, \lambda_s)$; thus $\beta_1 \neq 0$. Substituting λ_i for ω_i in (6.5) we get

$$f\left(\sum_j \alpha_{1j} \otimes x_{1j}, \ldots, \sum_j \alpha_{nj} \otimes x_{nj}\right) = \sum_k \beta_k \otimes a_k.$$

Since $\alpha_{ij}, \beta_k \in F$, we have $\sum_j \alpha_{ij} \otimes x_{ij} = 1 \otimes \left(\sum_j \alpha_{ij} x_{ij}\right)$ and $\sum_k \beta_k \otimes a_k = 1 \otimes \left(\sum_k \beta_k a_k\right)$. The above identity can be therefore rewritten as

$$1 \otimes f\left(\sum_j \alpha_{1j} x_{1j}, \ldots, \sum_j \alpha_{nj} x_{nj}\right) = 1 \otimes \left(\sum_k \beta_k a_k\right),$$

which is a contradiction. Namely, the left-hand side is 0 since f is an identity of A, while the right-hand side is not 0 since $\beta_1 \neq 0$ and the a_k's are linearly independent. $\qquad \square$

Theorem 6.29 can be proved in a more direct manner, avoiding the polynomial algebra. But we wanted to point out the general principle. If one wishes to prove a result for general commutative algebras, then it is sometimes enough to prove it for polynomial algebras. This approach will be used a few more times in the rest of the chapter.

Returning to scalar extensions, we now see that, as long as F is an infinite field, an F-algebra A and its scalar extension A_K satisfy the same identities (with coefficients in F). Indeed, an identity of A_K is also an identity of A since we can embed A into A_K via $x \mapsto 1 \otimes x$, and an identity of A is also an identity of A_K by Theorem 6.29. For instance, from Example 2.72 we now see that $\mathbb{H}$ and $M_2(\mathbb{C})$ satisfy the same identities (with real coefficients). More generally, by making use of splitting fields, the problem of determining identities of a finite dimensional central simple F-algebra A can be reduced to the case where $A = M_n(K)$. This is true even when F is finite since then A is already isomorphic to $M_n(K)$ by Corollary 2.71.

6.7 T-ideals

Throughout this section, X denotes the countably infinite set $\{\xi_1, \xi_2, \dots\}$ of non-commuting indeterminates. By $\mathrm{Id}(A)$ we denote the set of all polynomial identities of an F-algebra A. Obviously, $\mathrm{Id}(A)$ is an ideal of the free algebra $F\langle X\rangle$. One can say more about it, as we shall soon see. We restrict our discussion to nonzero unital algebras, so that evaluations make sense also for polynomials with nonzero constant term. This will simplify the exposition.

Definition 6.30 An ideal I of $F\langle X\rangle$ is called a **T-ideal** if $\varphi(I) \subseteq I$ for every algebra endomorphism φ of $F\langle X\rangle$.

Remark 6.31 If φ is an endomorphism of $F\langle X\rangle$ and $g_i := \varphi(\xi_i)$, $i \in \mathbb{N}$, then

$$\varphi\big(f(\xi_1, \dots, \xi_n)\big) = f(g_1, \dots, g_n)$$

for every $f \in F\langle X\rangle$. On the other hand, for any $g_i \in F\langle X\rangle$, $i \in \mathbb{N}$, there exists an endomorphism φ of $F\langle X\rangle$ satisfying $\varphi(\xi_i) = g_i$. Thus, an ideal I of $F\langle X\rangle$ is a T-ideal if and only if for all $f = f(\xi_1, \dots, \xi_n) \in I$ and $g_1, \dots, g_n \in F\langle X\rangle$ we have $f(g_1, \dots, g_n) \in I$.

If A is a nonzero unital algebra, then $\mathrm{Id}(A)$ is obviously a proper T-ideal of $F\langle X\rangle$. The next lemma in particular shows that the converse is also true. That is, for every proper T-ideal I of $F\langle X\rangle$ there exists a nonzero unital F-algebra A such that $I = \mathrm{Id}(A)$.

Lemma 6.32 *If I is a proper T-ideal of $F\langle X\rangle$, then $I = \mathrm{Id}(F\langle X\rangle/I)$.*

Proof Take an arbitrary polynomial $f = f(\xi_1, \dots, \xi_n) \in F\langle X\rangle$, and consider its evaluations on $F\langle X\rangle/I$. Note that

$$f(g_1 + I, \dots, g_n + I) = f(g_1, \dots, g_n) + I$$

for all $g_1, \dots, g_n \in F\langle X\rangle$. Using Remark 6.31 we thus see that $f \in I$ if and only if f is an identity of $F\langle X\rangle/I$.   □

Assume temporarily that I is just a nonempty subset of $F\langle X\rangle$. The **variety** determined by I, denoted by $\mathscr{V}(I)$, is the class of all unital algebras A such that $I \subseteq \mathrm{Id}(A)$; that is, every polynomial from I is an identity of A. For example, $\mathscr{V}\big(\{\xi_1\xi_2 - \xi_2\xi_1\}\big)$ is the class of all commutative unital algebras. Note that the variety determined by the set I coincides with the variety determined by the T-ideal generated by I. Therefore, assume from now on that I itself is a T-ideal; to avoid the trivial situation, assume also that $I \neq F\langle X\rangle$. Lemma 6.32 shows that the algebra $F\langle X\rangle/I$ belongs to $\mathscr{V}(I)$. Moreover, $F\langle X\rangle/I$ is a distinguished representative of the class $\mathscr{V}(I)$ due to the following property. If $A \in \mathscr{V}(I)$ and $\{a_1, a_2, \dots\}$ is a countably infinite subset of A, then there exists a unique unital homomorphism $\varphi : F\langle X\rangle/I \to A$ such that $\varphi(\xi_i + I) = a_i$, $i = 1, 2, \dots$. Indeed, just take the unital homomorphism from $F\langle X\rangle$

to A determined by $\xi_i \mapsto a_i$, note that I lies in its kernel since $A \in \mathcal{V}(I)$, and the existence of φ follows. The uniqueness is obvious. Because of this property, we call $F\langle X \rangle / I$ the **relatively free algebra** (with respect to I).

To each unital algebra A we can assign the T-ideal $\mathrm{Id}(A)$ and the corresponding relatively free algebra $F\langle X \rangle / \mathrm{Id}(A)$.

Example 6.33 If A is not a PI-algebra, then $\mathrm{Id}(A) = \{0\}$ and so the corresponding relatively free algebra is just the free algebra $F\langle X \rangle$.

Example 6.34 Let $A = M_n(F)$. We begin by introducing a seemingly unrelated notion. Let $\Omega = \left\{ \omega_{jk}^{(i)} \mid j, k = 1, \ldots, n, \ i = 1, 2, \ldots \right\}$ be a set of commuting indeterminates. For each i we set

$$g^{(i)} = \begin{bmatrix} \omega_{11}^{(i)} & \omega_{12}^{(i)} & \cdots & \omega_{1n}^{(i)} \\ \omega_{21}^{(i)} & \omega_{22}^{(i)} & \cdots & \omega_{2n}^{(i)} \\ \vdots & \vdots & \ddots & \vdots \\ \omega_{n1}^{(i)} & \omega_{n2}^{(i)} & \cdots & \omega_{nn}^{(i)} \end{bmatrix} \in M_n\big(F[\Omega]\big).$$

We call $g^{(i)}$ an $n \times n$ **generic matrix**. Thus, the entries of a generic matrix are distinct (independent) commuting indeterminates. The subalgebra of $M_n(F[\Omega])$ generated by all $n \times n$ generic matrices $g^{(i)}$, $i \in \mathbb{N}$, and the identity matrix is called the **algebra of $n \times n$ generic matrices**. It will be denoted by $\mathrm{GM}_n(F)$. Assume from now on that F is an *infinite field*. We claim that

$$F\langle X \rangle / \mathrm{Id}(M_n(F)) \cong \mathrm{GM}_n(F),$$

i.e., the **relatively free algebra** corresponding to $M_n(F)$ is isomorphic to the algebra of $n \times n$ generic matrices. To prove this, take an identity f of $M_n(F)$. Since F is infinite, Theorem 6.29 tells us that f is also an identity of $F[\Omega] \otimes M_n(F) \cong M_n\big(F[\Omega]\big)$ (cf. Example 4.22). But then f is also an identity of the subalgebra $\mathrm{GM}_n(F)$ of $M_n\big(F[\Omega]\big)$. Accordingly, $\mathrm{GM}_n(F) \in \mathcal{V}(I)$ where $I = \mathrm{Id}(M_n(F))$. As observed above, this implies that there exists a unital homomorphism $\varphi : F\langle X \rangle / I \to \mathrm{GM}_n(F)$ such that $\varphi(\xi_i + I) = g^{(i)}$, $i = 1, 2, \ldots$. It is clear that φ is surjective. Let us prove that it is also injective. Take $f = f(\xi_1, \ldots, \xi_n)$ such that $\varphi(f + I) = 0$, i.e., $f\big(g^{(1)}, \ldots, g^{(n)}\big) = 0$. Since the entries of the matrices $g^{(i)}$ are independent commuting indeterminates, we can replace them by arbitrary elements in F (cf. the previous section). This shows that $f \in I$. Therefore f is injective, and hence an isomorphism.

6.8 The Characteristic Polynomial

This section may appear as an intruder in the chapter. The main reason for its inclusion is that it is needed in the next section for the proof of the Amitsur-Levitzki Theorem. On the other hand, although the results of this section do not directly involve polynomial identities, they consider some other kind of identities that have many interactions with them. At any rate, the topics of this section are of general interest and importance.

Some of the readers may have already met what we are about to say in a linear algebra course, yet probably at a lower level of generality. A typical course considers only matrices with entries from a field. Many concepts, however, still make sense if a field is replaced by a more general ring. Say, the *determinant* of a matrix $A \in M_n(C)$, which we denote by $\det(A)$, can be defined in the usual way as long as C *is a commutative ring*. Accordingly, we can define the **characteristic polynomial** of A as

$$p_A(\omega) := \det(A - \omega I) \in C[\omega].$$

For any polynomial $q \in C[\omega]$ we define $q(A)$ in the same way as when C is a field. The **Cayley-Hamilton Theorem** says that

$$p_A(A) = 0.$$

Quite possibly the reader is familiar with this theorem only in the case where C is a field. However, from many of its proofs it is evident that C can be any commutative ring. On the other hand, the Cayley-Hamilton theorem for fields easily implies the Cayley-Hamilton Theorem for commutative rings. Indeed, assuming the truth of the theorem for fields, it follows that it is also true for commutative domains since they can be embedded into their fields of quotients. In particular, the theorem holds for the ring $\mathbb{Z}[\omega_{11}, \omega_{12}, \ldots, \omega_{nn}]$. Thus, if $g = (\omega_{ij})$ is a generic matrix, then $p_g(g) = 0$. Now take any commutative ring C, choose $c_{ij} \in C$, $1 \le i, j \le n$, substitute c_{ij} for ω_{ij} in this identity, and the validity of the Cayley-Hamilton Theorem for C follows.

For any matrix $A \in M_n(C)$ we define $\mathrm{tr}(A)$, the *trace* of A, in the usual way. Our goal is to express the coefficients of p_A through the traces of the powers of A. To this end, we need formulas on symmetric polynomials in commuting indeterminates with integer coefficients, which were discovered already by Isaac Newton. Recall that a polynomial is said to be *symmetric* if it remains unchanged after permuting the indeterminates. Important examples are the so-called **elementary symmetric polynomials**, defined by $e_0 := 1$ and

$$e_k = e_k(\omega_1, \ldots, \omega_n) := \sum_{1 \le i_1 < i_2 < \cdots < i_k \le n} \omega_{i_1} \omega_{i_2} \ldots \omega_{i_k},$$

$k = 1, \ldots, n$, i.e., e_k is the sum of all distinct products of k distinct indeterminates; in particular, e_1 is the sum of all ω_i's, and e_n is their product. Let ω be another indeterminate. Note that

$$(-1)^n(\omega - \omega_1)(\omega - \omega_2)\ldots(\omega - \omega_n) = \sum_{j=0}^{n}(-1)^j e_{n-j}\omega^j. \tag{6.6}$$

Elementary symmetric polynomials are thus intimately connected with the roots of polynomials. Another reason for their importance is that they generate the algebra of all symmetric polynomials, but we will not need this fact. Our goal is to connect the elementary symmetric polynomials with the symmetric polynomials

$$p_j = p_j(\omega_1, \ldots, \omega_n) := \sum_{i=1}^{n} \omega_i^j, \quad j \geq 1.$$

One can verify that

$$e_1 = p_1, \quad 2e_2 = e_1 p_1 - p_2, \quad 3e_3 = e_2 p_1 - e_1 p_2 + p_3, \tag{6.7}$$

which indicates the general rule.

Lemma 6.35 (Newton's formulas) *For $k = 1, \ldots, n$, we have*

$$k e_k = \sum_{j=1}^{k} (-1)^{j-1} e_{k-j} p_j. \tag{6.8}$$

Proof From (6.6) it follows that

$$\sum_{j=0}^{n} (-1)^j e_{n-j} \omega_i^j = 0, \quad i = 1, \ldots, n.$$

Summing up over all i we obtain (6.8) for $k = n$. The case where $k < n$ follows easily from this one, basically we are facing only a problem in notation. The one that we are using suggests that n is fixed and k varies. While this setting is natural in view of applications, in the proof that follows it is inconvenient to regard n as fixed. Let us therefore write, just for the purpose of this proof, $e_{k,n}$ for e_k and $p_{j,n}$ for p_j. Our goal is to show that the polynomial

$$P := k e_{k,n} - \sum_{j=1}^{k} (-1)^{j-1} e_{k-j,n} p_{j,n}$$

is 0. Note that for any $i, j \leq k$ we have

$$e_{i,n}(\omega_1, \ldots, \omega_k, 0, \ldots, 0) = e_{i,k}(\omega_1, \ldots, \omega_k),$$

$$p_{j,n}(\omega_1, \ldots, \omega_k, 0, \ldots, 0) = p_{j,k}(\omega_1, \ldots, \omega_k).$$

Since (6.8) holds for $n = k$, it follows that $P(\omega_1, \ldots, \omega_k, 0, \ldots, 0) = 0$. This means that P does not contain nonzero monomials in $\omega_1, \ldots, \omega_k$. Similarly, by setting zeros at other places, we see that P does not contain nonzero monomials in any set of k indeterminates. However, from the definition of P it is obvious that its nonzero monomials could only be of degree k, and therefore cannot involve more than k indeterminates. Hence $P = 0$. $\qquad\qquad\qquad\qquad\qquad\qquad\qquad\qquad\qquad\qquad\qquad$ $\square$

In the characteristic zero case we can therefore express the e_k's through the p_j's. Say, from (6.7) we obtain

$$e_1 = p_1, \quad e_2 = \frac{1}{2}\left(p_1^2 - p_2\right), \quad e_3 = \frac{1}{6}\left(p_1^3 - 3p_1p_2 + 2p_3\right).$$

In general, for each $k = 1, \ldots, n$ there exists a polynomial $q_k = q_k(\omega_1, \ldots, \omega_k) \in \mathbb{Q}[\omega_1, \omega_2, \ldots]$ such that

$$e_k = q_k(p_1, \ldots, p_k). \tag{6.9}$$

For our goals it is important to note that q_k has zero constant term. We also remark that if $\lambda\omega_1^{m_1} \ldots \omega_k^{m_k}, \lambda \in \mathbb{Q}, m_j \geq 0$, is a monomial of q_k, then $m_1 + 2m_2 + \cdots + km_k = k$. For convenience we also set $q_0 := 1$, so that (6.9) holds for $k = 0, 1, \ldots, n$.

Theorem 6.36 *If C is a commutative $\mathbb{Q}$-algebra, then for every $A \in M_n(C)$ we have*

$$p_A(\omega) = \sum_{j=0}^{n}(-1)^j q_{n-j}\big(\mathrm{tr}(A), \mathrm{tr}(A^2), \ldots, \mathrm{tr}(A^{n-j})\big)\omega^j.$$

Proof A standard reasoning shows that it suffices to consider the case where A is a generic matrix. Therefore we may assume that $C = \mathbb{Q}[\omega_{11}, \omega_{12} \ldots, \omega_{nn}]$. In particular, C is thus a commutative domain and as such it can be embedded into a field (e.g., its field of quotients), which can be further embedded into an algebraically closed fields (e.g., its algebraic closure). But then there is no loss of generality in assuming that C itself is an algebraically closed field. Hence we have $p_A(\omega) = (-1)^n(\omega - \lambda_1) \ldots (\omega - \lambda_n)$, where the λ_i's are the eigenvalues of A. From the Jordan normal form of A it readily follows that $\mathrm{tr}(A^i) = p_i(\lambda_1, \ldots, \lambda_n), i = 1, \ldots, n$. Using (6.6) we now obtain

$$p_A(\omega) = \sum_{j=0}^{n}(-1)^j e_{n-j}(\lambda_1, \ldots, \lambda_n)\omega^j$$

$$= \sum_{j=0}^{n}(-1)^j q_{n-j}\big(p_1(\lambda_1, \ldots, \lambda_n), \ldots, p_{n-j}(\lambda_1, \ldots, \lambda_n)\big)\omega^j$$

$$= \sum_{j=0}^{n}(-1)^j q_{n-j}\big(\mathrm{tr}(A), \ldots, \mathrm{tr}(A^{n-j})\big)\omega^j,$$

as desired. $\qquad\qquad\qquad\qquad\qquad\qquad\qquad\qquad\qquad\qquad\qquad\qquad\qquad$ $\square$

For example, for $n = 2$ we have

$$p_A(\omega) = \omega^2 - \text{tr}(A)\omega + \frac{1}{2}\left(\text{tr}(A)^2 - \text{tr}(A^2)\right) \qquad (6.10)$$

(and hence $\det(A) = \frac{1}{2}\left(\text{tr}(A)^2 - \text{tr}(A^2)\right)$ in this case). In general, the coefficient at ω^j is a $\mathbb{Q}$-linear combination of expressions of the form

$$\text{tr}(A)^{m_1} \text{tr}\left(A^2\right)^{m_2} \ldots \text{tr}\left(A^{n-j}\right)^{m_{n-j}}$$

where $m_1 + 2m_2 + \cdots + (n-j)m_{n-j} = n - j$.

Corollary 6.37 *Let C be a commutative $\mathbb{Q}$-algebra. If $A \in M_n(C)$ is such that* $\text{tr}(A) = \text{tr}(A^2) = \cdots = \text{tr}(A^n) = 0$, *then $A^n = 0$.*

Proof Since the polynomials q_k, $k = 1, \ldots, n$, have zero constant terms, it follows that $p_A(\omega) = (-1)^n \omega^n$. □

6.9 The Amitsur-Levitzki Theorem

The matrix algebra $M_n(F)$ has dimension n^2, and thus satisfies s_{n^2+1} (see Example 6.15). From (6.3) we therefore see that it also satisfies s_{n^2+2}, s_{n^2+3}, etc. But does it satisfy s_k with $k < n^2 + 1$? Moreover, what is the minimal degree of a nonzero identity of $M_n(F)$? We first approach these questions from the opposite direction.

Lemma 6.38 *A nonzero polynomial of degree less than $2n$ is not an identity of $M_n(F)$.*

Proof By Theorem 6.24 it suffices to treat multilinear polynomials. Take a nonzero multilinear polynomial f of degree $2n - 1$. Thus, $f = \sum_{\sigma \in S_{2n-1}} \lambda_\sigma \xi_{\sigma(1)} \cdots \xi_{\sigma(2n-1)}$ with, say, $\lambda_1 \neq 0$. The sequence of matrix units $E_{11}, E_{12}, E_{22}, \ldots, E_{n-1,n}, E_{nn}$ has the property that their product in the given order is E_{1n}, while the product in any other order is zero. Therefore $f(E_{11}, E_{12}, \ldots, E_{nn}) = \lambda_1 E_{1n} \neq 0$. Similarly, by taking an appropriate subsequence we see that a nonzero multilinear polynomial of degree less than $2n - 1$ is not an identity of $M_n(F)$. □

Thus, the best we can hope for is that $M_n(F)$ satisfies s_{2n}—and this is indeed the case! This was proved by S. A. Amitsur and J. Levitzki in 1950. The charming proof that we are about to give is due to S. Rosset.

Theorem 6.39 (Amitsur-Levitzki) *The standard polynomial s_{2n} is an identity of $M_n(C)$ for every commutative algebra C.*

Proof By assumption, C is an algebra over some field F. Let F_0 be the prime subfield of F, and consider C as an algebra over F_0. Since $M_n(C) \cong C \otimes_{F_0} M_n(F_0)$ (Example 4.22), and since multilinear identities are stable (Lemma 6.27), it is enough to prove that s_{2n} is an identity of $M_n(F_0)$. Suppose this is true for $F_0 = \mathbb{Q}$. Then s_{2n} is also an identity of $M_n(\mathbb{Z})$. But $M_n(\mathbb{Z}_p)$ is a homomorphic image of $M_n(\mathbb{Z})$. Indeed, if φ is the canonical homomorphism from $\mathbb{Z}$ onto $\mathbb{Z}_p$, then $(m_{ij}) \mapsto (\varphi(m_{ij}))$ is a homomorphism from $M_n(\mathbb{Z})$ onto $M_n(\mathbb{Z}_p)$. Therefore s_{2n} is also an identity of $M_n(\mathbb{Z}_p)$. We have thus reduced the problem to proving that s_{2n} is an identity of $M_n(\mathbb{Q})$.

Take $A_1, \ldots, A_{2n} \in M_n(\mathbb{Q})$. We must show that $s_{2n}(A_1, A_2, \ldots, A_{2n}) = 0$. We will accomplish this with the help of the Grassmann algebra $G = G_0 \oplus G_1$ over $\mathbb{Q}$ with generators $x_1, x_2, \ldots$ (see Example 6.7). For $B = (b_{ij}) \in M_n(\mathbb{Q})$ and $x \in G$ we define Bx to be the matrix $(b_{ij}x) \in M_n(G)$. Note that the product $(Bx)(B'x')$ of two such matrices is equal to $(BB')(xx')$. Let us set

$$A := A_1 x_1 + A_2 x_2 + \cdots + A_{2n} x_{2n} \in M_n(G).$$

From $x_i G x_i = 0$ we easily infer that

$$A^{2n} = \sum_{\sigma \in S_{2n}} A_{\sigma(1)} A_{\sigma(2)} \ldots A_{\sigma(2n)} x_{\sigma(1)} x_{\sigma(2)} \ldots x_{\sigma(2n)}.$$

Recall that $x_{\sigma(1)} \ldots x_{\sigma(r)} = \operatorname{sgn}(\sigma) x_1 \ldots x_r$ (see (6.4)). Consequently,

$$A^{2n} = s_{2n}(A_1, A_2, \ldots, A_{2n}) x_1 x_2 \ldots x_{2n}.$$

Hence we are reduced to proving that $A^{2n} = 0$, i.e., $(A^2)^n = 0$. Note that A^2 lies in $M_n(G_0)$. As G_0 is a commutative algebra, Corollary 6.37 tells us that it suffices to show that $(A^2)^k = A^{2k}$ has zero trace for $k = 1, \ldots, n$. To this end, notice that $A^{2k-1} = \sum_j B_j y_j$ for some $B_j \in M_n(\mathbb{Q})$ and $y_j \in G_1$. Consequently,

$$\operatorname{tr}(A^{2k}) = \operatorname{tr}(A A^{2k-1}) = \operatorname{tr}\left(\sum_{i,j} A_i B_j x_i y_j\right) = \sum_{i,j} \operatorname{tr}(A_i B_j) x_i y_j.$$

On the other hand,

$$\operatorname{tr}(A^{2k}) = \operatorname{tr}(A^{2k-1} A) = \sum_{i,j} \operatorname{tr}(B_j A_i) y_j x_i.$$

Comparing both expressions and using $\operatorname{tr}(A_i B_j) = \operatorname{tr}(B_j A_i)$ (a general property of the trace) and $x_i y_j = -y_j x_i$ (a consequence of $y_j \in G_1$), we obtain $\operatorname{tr}(A^{2k}) = 0$. $\qquad \square$

Exercises

6.1. Suppose that $f \in F\langle \xi, \eta \rangle$ commutes with $\xi\eta$. Show that there exists $p \in F[\omega]$ such that $f = p(\xi\eta)$.

Remark: This is a simple special case of a theorem by G. Bergman: If $f, g \in F\langle X \rangle$ commute, then there exist $h \in F\langle X \rangle$ and $p, q \in F[\omega]$ such that $f = p(h)$ and $g = q(h)$.

6.2. Find an automorphism of $F\langle \xi, \eta \rangle$ different from the identity map.

6.3. Show that $F\langle \xi_1, \xi_2, \dots \rangle$ can be embedded into $F\langle \xi, \eta \rangle$ via $\xi_i \mapsto \xi^i \eta$.

6.4. A linear subspace J of an algebra A is called a **Jordan subalgebra** of A if $u \circ v := uv + vu \in J$ for all $u, v \in J$. For example, the set of symmetric matrices in $M_n(F)$, $n \geq 2$, is a Jordan subalgebra of $M_n(F)$, but not a subalgebra. Elements of the Jordan subalgebra of $F\langle X \rangle$ generated by $X \cup \{1\}$ are called **Jordan polynomials**. For example, if $\text{char}(F) \neq 2$ then $\xi_1^2 = \frac{1}{2}\xi_1 \circ \xi_1$ and $\xi_1\xi_2\xi_1 = \frac{1}{2}\xi_1 \circ (\xi_1 \circ \xi_2) - \frac{1}{4}(\xi_1 \circ \xi_1) \circ \xi_2$ are Jordan polynomials. Show that $\xi_1^n, \xi_1\xi_2\xi_3 + \xi_3\xi_2\xi_1, (\xi_1\xi_2 - \xi_2\xi_1)^2$ are also Jordan polynomials. Find a Jordan subalgebra J of the Grassmann algebra G such that $u_1u_2u_3u_4 + u_4u_3u_2u_1 \notin J$ for some $u_1, u_2, u_3, u_4 \in J$. Hence, conclude that $\xi_1\xi_2\xi_3\xi_4 + \xi_4\xi_3\xi_2\xi_1$ is not a Jordan polynomial, and neither is $\xi_1\xi_2 \dots \xi_n + \xi_n\xi_{n-1} \dots \xi_1$ for any $n \geq 4$.

Remark: A nonassociative ($=$ not necessarily associative) algebra J is called a **Jordan algebra** if $xy = yx$ and $x^2(yx) = (x^2y)x$ for all $x, y \in J$. Every Jordan subalgebra of an associative algebra is a Jordan algebra under the product $\circ$. Such Jordan algebras are called **special**. There exist Jordan algebras that are not special; they are called **exceptional**.

6.5. Analogously, a **Lie subalgebra** of an algebra A is defined as a linear subspace of A, say L, such that $[u, v] \in L$ for all $u, v \in L$. For example, the set of all trace zero matrices in $M_n(F)$, $n \geq 2$, is a Lie subalgebra of $M_n(F)$, but not a subalgebra. A **Lie polynomial** is an element of the Lie subalgebra of $F\langle X \rangle$ generated by X. Find a multilinear Lie polynomial of degree 3 which is simultaneously a Jordan polynomial. Show that $\xi_1\xi_2 \dots \xi_n - \xi_n\xi_{n-1} \dots \xi_1$ is not a Lie polynomial for $n \geq 3$.

Remark: A nonassociative algebra L is called a **Lie algebra** if $x^2 = 0$ and $(xy)z + (zx)y + (yz)x = 0$ for all $x, y, z \in L$. A Lie subalgebra of an associative algebra is a Lie algebra under the product $[.\,,.]$. Conversely, the **Poincaré-Birkhoff-Witt Theorem** shows that every Lie algebra is isomorphic to a Lie subalgebra of an associative algebra. (The product in Lie algebras is actually usually denoted by $[.\,,.]$, so the defining identities are written as $[x, x] = 0$ and $[[x, y], z] + [[z, x], y] + [[y, z], x] = 0$; the latter is called the **Jacobi identity**.)

6.6. Describe the $\mathbb{R}$-algebras $\mathbb{C}$, $\mathbb{R} \times \mathbb{R}$, and $\mathbb{H}$ in terms of generators and relations.

6.7. Find all invertible elements in $F\langle x, x' \mid xx' = x'x = 1\rangle$.

Remark: A more standard notation for this algebra is $F[x, x^{-1}]$. Its elements are called **Laurent polynomials** (cf. Example 1.41). Note also that $F[x, x^{-1}]$ is isomorphic to the group algebra $F[\mathbb{Z}]$.

6.8. Let $q \in F^*$. Show that $F\langle x, x', y, y' \mid xx' = x'x = 1, yy' = y'y = 1, xy = qyx\rangle$ is a simple noetherian domain provided that q is not a root of unity.

6.9. Show that $F\langle x, y \mid xy = 1\rangle$ is a primitive algebra with nonzero socle, and that it contains a left zero-divisor which is not a right zero-divisor.

Hint: Examine the element $e := 1 - yx$.

6.10. Prove that $F\langle \xi_1, \xi_2\rangle \otimes F\langle \eta_1, \eta_2\rangle \cong F\langle x_1, x_2, y_1, y_2 \mid [x_i, y_j] = 0, i, j = 1, 2\rangle$.

6.11. Prove that the Weyl algebras satisfy $\mathscr{A}_n \cong \mathscr{A}_k \otimes \mathscr{A}_{n-k}$, $1 \leq k < n$.

6.12. Find the Jacobson radical of the Grassmann algebra G.

6.13. Let $f_i(\xi_1, \ldots, \xi_n) \in F\langle X\rangle$, $i = 1, \ldots, m$, be linearly independent polynomials. Show that an F-algebra A must be a PI-algebra if for arbitrary $x_1, \ldots, x_n \in A$ the evaluations $f_i(x_1, \ldots, x_n)$, $i = 1, \ldots, m$, are linearly dependent in A.

6.14. We say that an F-algebra A is an **algebraic algebra of bounded degree** if there exists $n \in \mathbb{N}$ such that for every $x \in A$ there is $f(\omega) \in F[\omega]$ of degree n satisfying $f(x) = 0$. Show that the class of algebraic algebras of bounded degree properly contains the class of finite dimensional algebras, and is properly contained in the class of PI-algebras.

6.15. Show that the algebra of all finite rank operators on an infinite dimensional vector space over F is an algebraic algebra, but not a PI-algebra.

6.16. Let A be an algebra and let M be an A-bimodule. Equip $A \times M$ with pointwise addition and scalar multiplication, and the product $(x, m)(y, n) = (xy, xn+my)$. Check that in this way $A \times M$ becomes an algebra; let us call it A_M. Show that A is a PI-algebra if and only if A_M is a PI-algebra.

6.17. Find an identity of minimal degree of the subalgebra of $M_n(F)$ consisting of all matrices that have zeros in all columns except in the first one.

6.18. Show that, in general, $\mathrm{Id}(A) \neq \mathrm{Id}(A[\omega])$, although A and $A[\omega]$ satisfy the same multilinear identities.

6.19. Let $f = f(\xi_1, \ldots, \xi_n) \in F\langle X\rangle$ be a multilinear polynomial, and let $k \in \mathbb{N}$. Prove that $\mathrm{tr}\big(f(A_1, \ldots, A_n)\big) = 0$ for all $A_1, \ldots, A_n \in M_k(F)$ if and only if there exist $f_0 \in \mathrm{Id}(M_k(F))$ and $g_i, h_i \in F\langle X\rangle$ such that $f = f_0 + \sum_i [g_i, h_i]$.

Hint: Use $m\xi_n m' = m'm\xi_n + [m\xi_n, m']$ for monomials m, m'.

6.20. Show that every generic matrix in $M_n(F[\Omega])$ is invertible in $M_n(F(\Omega))$. Hence conclude that if F is an infinite field, $m \in F\langle X\rangle$ is a nonzero monomial, and $f \in F\langle X\rangle$ is such that $mf \in \mathrm{Id}(M_n(F))$, then $f \in \mathrm{Id}(M_n(F))$.

Remark: A theorem by Amitsur says that this is true even when m is an arbitrary polynomial which is not an identity of $M_n(F)$. See also the comment following Theorem 7.65.

6.21. Let A be a nonzero finite dimensional semiprime algebra. Show that there exists $n \in \mathbb{N}$ such that A satisfies s_{2n}, but does not satisfy nonzero identities of degree less than $2n$.

6.22. Given $A, B \in M_2(F)$, let

$$P_{A,B}(\xi, \eta) := \xi\eta + \eta\xi - \text{tr}(B)\xi - \text{tr}(A)\eta + \text{tr}(A)\text{tr}(B) - \text{tr}(AB) \in F\langle\xi, \eta\rangle.$$

Recalling the form (6.10) of the characteristic polynomial, we see that by applying the linearization method to the Cayley-Hamilton Theorem we obtain $P_{A,B}(A, B) = 0$. Show that the identity

$$\sum_{\sigma \in S_4} \text{sgn}(\sigma)P_{A_{\sigma(1)}A_{\sigma(2)}, A_{\sigma(3)}A_{\sigma(4)}} \left(A_{\sigma(1)}A_{\sigma(2)}, A_{\sigma(3)}A_{\sigma(4)}\right) = 0$$

yields the Amitsur-Levitzki Theorem for $n = 2$ (provided that $\text{char}(F) \neq 2$, but from the first paragraph of our proof of the Amitsur-Levitzki Theorem it is evident that this can be assumed without loss of generality).

Remark: We have confined ourselves to $n = 2$ for clarity. The Amitsur-Levitzki Theorem can be, in a similar fashion, derived from the Cayley-Hamilton theorem for any n. This was discovered by Yu. P. Razmyslov, and is actually one of the standard proofs of the Amitsur-Levitzki Theorem. A deeper result, obtained independently by Razmyslov and C. Procesi, shows that in fact *every* polynomial identity of $M_n(F)$, $\text{char}(F) = 0$, is a consequence of the Cayley-Hamilton identity $p_A(A) = 0$. Moreover, the same is true for the more general **trace identities**. We omit giving their definition, but just give a hint by saying that the characteristic polynomial p_A gives rise to a trace identity that is not a polynomial identity.

Chapter 7
Rings of Quotients and Structure of PI-Rings

A commutative domain can be embedded into a field, its field of quotients. One usually learns this simple and highly useful fact in one's first encounters with abstract algebra. Therefore it is natural to seek its analogies in noncommutative rings. The lack of commutativity causes considerable difficulties. Nevertheless, various generalizations of the construction of the field of quotients exist in noncommutative ring theory. We will discuss a few of them: rings of central quotients, classical rings of quotients, and Martindale rings of quotients. Although now the issue is the noncommutativity, we cannot entirely escape from the commutative context. The centers of the constructed rings will be of crucial importance for us. The so-called extended centroid of a prime ring, i.e., a field defined as the center of the Martindale ring of quotients, will enable us to extend a part of the theory of central simple algebras to general prime rings. We will also use it as a principal tool in the second part of the chapter, devoted to the structure theory of prime rings satisfying polynomial identities (or their generalizations, the generalized polynomial identities). As a very rough summary, it can be stated that prime PI-rings are close to finite dimensional simple algebras.

7.1 Rings of Central Quotients

A glance at the construction of the field of quotients of a commutative domain shows that it highly depends on the commutativity. There is no obvious way to avoid it. However, in this first section we will take an easy way; we are going to use essentially the same construction for a noncommutative ring R, but allow only the elements from the center $Z = Z(R)$ as "denominators". More precisely, we will assume that the nonzero elements from Z satisfy the (obviously necessary) condition from the next definition, and enlarge R in such a way that these elements become invertible in the new, larger ring.

© Springer International Publishing Switzerland 2014
M. Brešar, *Introduction to Noncommutative Algebra*, Universitext,
DOI 10.1007/978-3-319-08693-4_7

Definition 7.1 A nonzero element a in a ring R is said to be **regular** if it is neither a left nor right zero-divisor.

Constructing the ring of central quotients $Q_Z(R)$. We assume that R is any ring such that its center Z is nonzero and all elements in $Z\backslash\{0\}$ are regular; in particular, Z is a commutative domain. We claim that the relation $\sim$ on the set $R \times (Z\backslash\{0\})$ defined by

$$(r, z) \sim (r', z') \iff rz' = r'z$$

is an equivalence relation. Only the transitivity is not entirely obvious. Thus, assume that $(r, z) \sim (r', z')$ and $(r', z') \sim (r'', z'')$, i.e., $rz' = r'z$ and $r'z'' = r''z'$. Then $(rz'' - r''z)z' = r'zz'' - r'z''z = 0$, and so $(r, z) \sim (r'', z'')$ follows from the regularity of z'. Let rz^{-1} denote the equivalence class of (r, z). Thus, $rz^{-1} = r'z'^{-1}$ if and only if $rz' = r'z$; in particular, $rz(zw)^{-1} = rw^{-1}$. Let $Q_Z(R)$ denote the set of all equivalence classes, equipped with addition and multiplication defined by

$$rz^{-1} + sw^{-1} := (rw + sz)(zw)^{-1},$$
$$rz^{-1} \cdot sw^{-1} := rs(zw)^{-1}.$$

To prove that these operations are well-defined, we must show that $(r, z) \sim (r', z')$ and $(s, w) \sim (s', w')$ imply $(rw + sz, zw) \sim (r'w' + s'z', z'w')$ and $(rs, zw) \sim (r's', z'w')$. But this is straightforward. As one would expect, $Q_Z(R)$ equipped with these operations is a ring. Verifying this requires a little patience, but no more than verifying that $\mathbb{Q}$ is a ring.

Definition 7.2 The ring $Q_Z(R)$ is called the **ring of central quotients** of R.

No matter whether R is unital or not, $Q_Z(R)$ is. Its unity is $z_0z_0^{-1}$, where z_0 is an arbitrary element in $Z\backslash\{0\}$. The center of $Q_Z(R)$, which we denote by $\widehat{Z}$, is easily seen to consist of elements of the form zw^{-1}, $z, w \in Z$, $w \neq 0$. If $z \neq 0$, then wz^{-1} is the inverse of zw^{-1}. Therefore $\widehat{Z}$ is a field; of course, $\widehat{Z}$ is nothing but the field of quotients of Z. Every ring whose center is a field can be considered as a central algebra over its center. Thus, $Q_Z(R)$ is a central algebra over $\widehat{Z}$.

One immediately checks that

$$\iota : R \to Q_Z(R), \quad \iota(r) = (rz_0)z_0^{-1},$$

is an embedding of R into $Q_Z(R)$ (we remark that the definition of ι is independent of the choice of $z_0 \in Z\backslash\{0\}$). It is convenient to identify R with its isomorphic copy $\iota(R)$ inside $Q_Z(R)$, and thus consider R as a subring of $Q_Z(R)$. The expression rz^{-1}, so far used just as a notation for the equivalence class, can then be interpreted as the product of $r \in R$ and the inverse of $z \in Z\backslash\{0\}$. Every element in $Z\backslash\{0\}$ is invertible in $Q_Z(R)$. Intuitively, $Q_Z(R)$ is the smallest ring containing R that has this property. The next proposition says this in more formal terms, through a *universal property*.

Proposition 7.3 *If T is a unital ring and $\varphi : R \to T$ is a homomorphism such that $\varphi(z)$ is invertible in T for every $z \in Z \backslash \{0\}$, then φ can be uniquely extended to a homomorphism $\overline{\varphi} : Q_Z(R) \to T$.*

Proof It is straightforward to check that $\overline{\varphi}(rz^{-1}) := \varphi(r)\varphi(z)^{-1}$ is a well-defined homomorphism which extends φ. The uniqueness is clear.   □

Let us add that if φ is injective and every element in T can be written as $\varphi(r)\varphi(z)^{-1}$, then $\overline{\varphi}$ is an isomorphism.

The next theorem summarizes our discussion up to now.

Theorem 7.4 *Let R be a ring such that its center Z is nonzero and all elements in $Z \backslash \{0\}$ are regular. The ring $Q_Z(R)$ has the following properties:*

(a) *$Q_Z(R)$ is a unital ring containing R as a subring.*
(b) *Every element in $Z \backslash \{0\}$ is invertible in $Q_Z(R)$.*
(c) *Every element in $Q_Z(R)$ is of the form rz^{-1}, where $r \in R$ and $z \in Z \backslash \{0\}$.*

Moreover, these properties characterize $Q_Z(R)$ up to isomorphism.

It is easy to see that if R is a prime ring with nonzero center Z, then $Q_Z(R)$ exists (i.e., nonzero elements in Z are regular) and is a prime ring.

We continue with examples. The first two are obvious, but worth mentioning.

Example 7.5 If the center of a unital ring R is a field, then $Q_Z(R) = R$.

Example 7.6 If R is a commutative domain, then $Q_Z(R)$ is the field of quotients of R.

Example 7.7 We claim that $Q_Z(M_n(\mathbb{Z})) \cong M_n(\mathbb{Q})$. Obviously, $M_n(\mathbb{Q})$ has the property (a). Commuting with matrix units we easily notice that the center of $M_n(\mathbb{Z})$ consists of matrices nI with $n \in \mathbb{Z}$ (cf. the proof of Lemma 1.15). Therefore $M_n(\mathbb{Q})$ has the property (b), as well as (c); to notice the latter, make use of the common denominator of the entries of a matrix in $M_n(\mathbb{Q})$.

Example 7.8 A small variation of the preceding example: If $R = \left[\begin{smallmatrix} \mathbb{Z} & 2\mathbb{Z} \\ \mathbb{Z} & \mathbb{Z} \end{smallmatrix} \right]$, i.e., the subring of $M_2(\mathbb{Z})$ consisting of matrices whose $(1, 2)$ entry is an even integer, then $Q_Z(R) = M_2(\mathbb{Q})$. The point we wish to make here is that even if a ring is a bit "odd", its ring of central quotients may be "spotless".

Example 7.9 If R is the ring of quaternions with integer coefficients, $R = \mathbb{Z} \oplus \mathbb{Z}i \oplus \mathbb{Z}j \oplus \mathbb{Z}k$, then $Q_Z(R)$ is the ring of quaternions with rational coefficients, $Q_Z(R) = \mathbb{Q} \oplus \mathbb{Q}i \oplus \mathbb{Q}j \oplus \mathbb{Q}k$. Indeed, $Z(R) = \mathbb{Z}$ and checking (a)–(c) is immediate.

Certain simple modifications of our construction of $Q_Z(R)$ are also important and can be found in standard textbooks, usually under the name "(central) localization". We will not need them, but let us just give a clue what changes can be made. The role of $Z \backslash \{0\}$ can be replaced by any subset S of Z that is closed under multiplication, and instead of requiring that the elements from S are regular, one defines the equivalence relation $\sim$ on $R \times S$ as follows: $(r, z) \sim (r', z') \iff (rz' - r'z)z'' = 0$ for some $z'' \in S$.

7.2 Classical Rings of Quotients

Given a ring R, is it possible to construct a larger ring Q such that *all* regular elements in R, not only those from the center, are invertible in Q? More specifically, can we construct Q in such a way that Theorem 7.4 with the set of all regular elements playing the role of $Z\backslash\{0\}$ holds for Q? The answer is not as straightforward as for the ring of central quotients. It certainly is positive for some rings (for commutative domains, for example). But it can also be negative, as we will see.

Let us make it precise what ring Q we wish to have. To avoid trivialities, we tacitly assume that our rings contain at least one regular element.

Definition 7.10 Let R be a ring and S be the set of all regular elements in R. A ring Q is called a **right classical ring of quotients** of R if it has the following properties:

(a) Q is a unital ring containing R as a subring.
(b) Every element in S is invertible in Q.
(c) Every element in Q is of the form rs^{-1}, where $r \in R$ and $s \in S$.

It is clear that elements of R that are not regular cannot be invertible in any larger ring. The existence of a right classical ring of quotients of R is thus the best we can hope for if we wish to make as many elements in R invertible as possible.

A **left** classical ring of quotients of R is defined analogously, one just substitutes $s^{-1}r$ for rs^{-1} in (c). However, up until the end of the section we will consider only right classical rings of quotients.

Example 7.11 Let R be any of the rings from Examples 7.6–7.9, i.e., a commutative domain, $M_n(\mathbb{Z})$, $\left[\begin{smallmatrix} \mathbb{Z} & 2\mathbb{Z} \\ \mathbb{Z} & \mathbb{Z} \end{smallmatrix}\right]$, or $\mathbb{Z} \oplus \mathbb{Z}i \oplus \mathbb{Z}j \oplus \mathbb{Z}k$. It is easy to verify that the ring of central quotients $Q_Z(R)$ is also a right classical ring of quotients of R.

Example 7.12 If every regular element in R is already invertible, then R is its own right classical ring of quotients. Every finite dimensional unital algebra thus has this property (see Remark 1.31 (b)).

Suppose R has a right classical ring of quotients Q. Given $r \in R$ and $s \in S$, we have $s^{-1}r \in Q$ by (a) and (b), and so by (c) there exist $r_1 \in R$ and $s_1 \in S$ such that $s^{-1}r = r_1 s_1^{-1}$. Multiplying by s from the left and by s_1 from the right we obtain $rs_1 = sr_1$. Thus, the following holds:

$$rS \cap sR \neq \emptyset \quad \text{for every } r \in R \text{ and } s \in S. \qquad (7.1)$$

It is easy to find rings in which (7.1) is not fulfilled, and hence rings not having right classical ring of quotients (see Example 7.18 below). The main goal of this section is to prove that (7.1) is also sufficient, not only necessary, condition for a ring R to have a right classical ring of quotients. This result is basically due to O. Ore; more precisely, he proved it, in 1931, for the special case where R is a domain. The condition (7.1) is called the **right Ore condition**.

Unlike in the previous section, the noncommutativity is now a serious obstacle. There are two standard ways to overcome it. The first one, used already by Ore, is based on a careful modification of the construction from the previous section. While this approach is intuitively clear, working out the details is extremely tedious. We will therefore choose the other way. To gain intuition, we sketch it for the seminal case $R = \mathbb{Z}$. Via the regular representation we may identify $\mathbb{Z}$ with the maps $x \mapsto nx$, $n \in \mathbb{Z}$, from $\mathbb{Z}$ into itself. Our goal is to describe $\mathbb{Q}$ without explicit use of fractions, i.e., by staying inside the framework of the ring $\mathbb{Z}$. Consider, for example, the map $f : 2\mathbb{Z} \to \mathbb{Z}, f(2x) = x$. Note that f halves each number from its domain, so it carries information about the rational number $\frac{1}{2}$. Similarly, $g : m\mathbb{Z} \to \mathbb{Z}, g(mx) = nx$, where $m \in \mathbb{N}$ and $n \in \mathbb{Z}$, corresponds to $\frac{n}{m} \in \mathbb{Q}$ (in the sense that $g(u) = \frac{n}{m}u$ for all $u \in m\mathbb{Z}$). Given another map of this kind, $g' : m'\mathbb{Z} \to \mathbb{Z}, g'(m'x) = n'x$, we can consider $g + g'$ and gg' as maps defined on $mm'\mathbb{Z}$ (maybe one can choose larger domains, but this is irrelevant for us). Note that $g + g'$ corresponds to $\frac{n}{m} + \frac{n'}{m'}$, and gg' to $\frac{n}{m} \cdot \frac{n'}{m'}$. By identifying maps that correspond to the same rational number we can obtain, in this way, an alternative definition of the field $\mathbb{Q}$. Perhaps an overcomplicated one, but having an advantage of being suitable for generalizations to noncommutative rings. The crucial point is that the above maps are $\mathbb{Z}$-module homomorphisms from ideals of the ring $\mathbb{Z}$ into $\mathbb{Z}$.

Constructing the right classical ring of quotients $Q_{rc}(R)$. Let R be a ring satisfying the right Ore condition (7.1). Denote by $\mathscr{I}$ the set of all right ideals of R that contain at least one regular element. A right ideal can be considered as a right R-module. Consider the set of all pairs (f, I) where $I \in \mathscr{I}$ and $f : I \to R$ is a right R-module homomorphism. Define an equivalence relation on this set as follows: $(f, I) \sim (g, J)$ if f and g coincide on some $K \in \mathscr{I}$ such that $K \subseteq I \cap J$. In order to verify the transitivity, it is enough to notice that $I_1 \cap I_2 \in \mathscr{I}$ if both $I_1, I_2 \in \mathscr{I}$. Let us therefore prove this. Take $s_i \in I_i \cap S, i = 1, 2$, and use (7.1) to find $s \in S$ and $r \in R$ such that $s_1 s = s_2 r$. Hence $I_1 \cap I_2$ contains a regular element, namely $s_1 s$, so it belongs to $\mathscr{I}$.

We write $[f, I]$ for the equivalence class determined by (f, I). Let $Q_{rc}(R)$ denote the set of all equivalence classes, endowed with addition and multiplication

$$[f_1, I_1] + [f_2, I_2] := [f_1 + f_2, I_1 \cap I_2],$$
$$[f_1, I_1][f_2, I_2] := [f_1 f_2, f_2^{-1}(I_1)].$$

We must prove that these operations are well-defined. First of all, $f_2^{-1}(I_1) = \{x \in I_2 \mid f_2(x) \in I_1\}$ indeed lies in $\mathscr{I}$. Namely, from (7.1) we infer that given $s_i \in I_i \cap S$, $i = 1, 2$, there exist $s \in S$ and $r \in R$ such that $f_2(s_2)s = s_1 r$, showing that the right ideal $f_2^{-1}(I_1)$ contains the regular element $s_2 s$. Assume now that $(f_1, I_1) \sim (g_1, J_1)$ and $(f_2, I_2) \sim (g_2, J_2)$. Thus, there exist $K_i \in \mathscr{I}$ such that $K_i \subseteq I_i \cap J_i$ and $f_i = g_i$ on K_i, $i = 1, 2$. Consequently, $f_1 + f_2 = g_1 + g_2$ on $K_1 \cap K_2 \in \mathscr{I}$ and hence $(f_1 + f_2, I_1 \cap I_2) \sim (g_1 + g_2, J_1 \cap J_2)$, which proves that addition is well-defined. Concerning multiplication, note that $f_1 f_2 = g_1 g_2$ on $f_2^{-1}(K_1) \cap K_2 \in \mathscr{I}$ and hence $(f_1 f_2, f_2^{-1}(I_1)) \sim (g_1 g_2, g_2^{-1}(J_1))$. Thus, both operations are indeed well-defined. Verifying that $Q_{rc}(R)$ is a ring with zero element $[0, R]$ and unity $[\mathrm{id}_R, R]$ is straightforward.

This construction is the basic example of the so-called "*noncommutative localization*". Its general version considers an arbitrary set closed under multiplication instead of the set of all regular elements. This does not make the construction substantially more complicated, but we will not go into this matter.

We are now in a position to prove the result announced above.

Theorem 7.13 (Ore) *A ring R has a right classical ring of quotients if and only if R satisfies the right Ore condition.*

Proof We only have to prove the "if" part. Assume that R satisfies the right Ore condition and let us show that $Q_{rc}(R)$ *is a right classical ring of quotients of R.*

We embed R into $Q_{rc}(R)$ via

$$\iota : R \to Q_{rc}(R), \quad \iota(r) = [L_r, R];$$

here, as always, L_r stands for a left multiplication map. Let us only check the injectivity. If $[L_r, R] = 0$, then $L_r = 0$ on some $I \in \mathscr{I}$. Since I contains a regular element, it follows that $r = 0$.

Let $s \in S$. Then the map $f_s : sR \to R, f_s(sx) = x$, is well-defined. Moreover, f_s is a right R-module homomorphism. Note that $sR \in \mathscr{I}$ since $s^2 \in sR \cap S$. The reader can readily verify that $[f_s, sR]$ is the inverse of $\iota(s)$. In particular, $\iota(s)$ is invertible.

Now take an arbitrary $[f, I] \in Q_{rc}(R)$. Pick $s \in I \cap S$ and set $r := f(s) \in R$. We have $f(sx) = rx = L_r f_s(sx)$ for all $x \in R$. Hence

$$[f, I] = [f, sR] = [L_r f_s, sR] = [L_r, R][f_s, sR] = \iota(r)\iota(s)^{-1}.$$

Identifying R with $\iota(R)$ we thus see that $Q_{rc}(R)$ has all the desired properties. $\square$

The ring $Q_{rc}(R)$ has a *universal property*, analogous to that satisfied by $Q_Z(R)$ (see Proposition 7.3).

Proposition 7.14 *Let R be a ring satisfying the right Ore condition. If T is a unital ring and $\varphi : R \to T$ is a homomorphism such that $\varphi(s)$ is invertible in T for every regular element $s \in R$, then φ can be uniquely extended to a homomorphism $\overline{\varphi} : Q_{rc}(R) \to T$.*

Proof Take $[f, I] \in Q_{rc}(R)$. Pick $s \in I \cap S$ and define $\overline{\varphi} : Q_{rc}(R) \to T$ by

$$\overline{\varphi}([f, I]) := \varphi(f(s))\varphi(s)^{-1}.$$

One can immediately check that $\overline{\varphi}(\iota(r)) = \varphi(r)$, so $\overline{\varphi}$ extends φ. However, we must first prove that $\overline{\varphi}$ is well-defined. Assume, therefore, that $(f, I) \sim (g, J)$. Our goal is to show that $\varphi(f(s))\varphi(s)^{-1} = \varphi(g(t))\varphi(t)^{-1}$ for every $t \in J \cap S$. Let $K \in \mathscr{I}$ be such that $K \subseteq I \cap J$ and $f = g$ on K. It suffices to show that

$\varphi(f(s))\varphi(s)^{-1} = \varphi(f(u))\varphi(u)^{-1}$ holds for $u \in K \cap S$. Namely, by analogy we then also have $\varphi(g(t))\varphi(t)^{-1} = \varphi(g(u))\varphi(u)^{-1}$, and so the desired conclusion follows from $f(u) = g(u)$. By regularity of s and u we have $sR, uR \in \mathcal{I}$, and hence $sR \cap uR \in \mathcal{I}$. Therefore there exist $x, y \in R$ such that $v := sx = uy \in S$. Since $\varphi(s)$ and $\varphi(v)$ are invertible in T by assumption, $v = sx$ gives $\varphi(s)^{-1} = \varphi(x)\varphi(v)^{-1}$. Consequently,

$$\varphi(f(s))\varphi(s)^{-1} = \varphi(f(s))\varphi(x)\varphi(v)^{-1} = \varphi(f(v))\varphi(v)^{-1}.$$

Similarly, from $v = uy$ we infer $\varphi(f(u))\varphi(u)^{-1} = \varphi(f(v))\varphi(v)^{-1}$. Comparing we get $\varphi(f(s))\varphi(s)^{-1} = \varphi(f(u))\varphi(u)^{-1}$.

The additivity of $\overline{\varphi}$ is clear. To prove the multiplicativity, take $[f_i, I_i] \in Q_{rc}(R)$, $i = 1, 2$. Pick $s_1 \in I_1 \cap S$. By definition we have $\overline{\varphi}([f_1, I_1][f_2, I_2]) = \varphi(f_1(f_2(s_2)))$ $\varphi(s_2)^{-1}$ where s_2 is an arbitrary element in $f_2^{-1}(I_1) \cap S$; since $[f_1, I_1] = [f_1, s_1 R]$ we may choose $s_2 \in f_2^{-1}(s_1 R)$. Thus $f_2(s_2) = s_1 x$ and hence $\varphi(x) = \varphi(s_1)^{-1}$ $\varphi(f_2(s_2))$, yielding

$$\varphi(f_1(f_2(s_2))) = \varphi(f_1(s_1))\varphi(x) = \varphi(f_1(s_1))\varphi(s_1)^{-1}\varphi(f_2(s_2)).$$

This readily implies that $\overline{\varphi}([f_1, I_1][f_2, I_2]) = \overline{\varphi}([f_1, I_1])\overline{\varphi}([f_2, I_2])$. Thus, $\overline{\varphi}$ is a homomorphism extending φ. The uniqueness is obvious because of condition (c) from Definition 7.10. □

Corollary 7.15 *If a ring satisfies the right Ore condition, then its right classical ring of quotients is unique up to isomorphism.*

Remark 7.16 Similar results of course hold for the *left* classical ring of quotients. One defines the **left Ore condition** as $Sr \cap Rs \neq \emptyset$ for every $r \in R$ and $s \in S$, and constructs the left classical ring of quotients $Q_{lc}(R)$ of R by using left ideals and right multiplication maps for defining the canonical embedding. Proposition 7.14 together with its left analogue easily implies that if R has both a left and right classical ring of quotients, then they are isomorphic. In other words, if both the left and right Ore condition are fulfilled, then $Q_{rc}(R) \cong Q_{lc}(R)$.

7.3 Ore Domains

In this section we take a closer look at the classical situation where R is a domain. The right Ore condition (7.1) in this case gets a slightly simpler form:

$$rR \cap sR \neq 0 \quad \text{for every } r, s \in R \setminus \{0\}. \tag{7.2}$$

Definition 7.17 A nonzero domain satisfying the right Ore condition (7.2) is called a **right Ore domain**.

Theorem 7.13 implies that a ring R is a right Ore domain if and only if R is a subring of a division ring D $(=Q_{rc}(R))$ such that every element in D is of the form rs^{-1} with $r, s \in R$, $s \neq 0$.

Not every domain is right Ore.

Example 7.18 Let $R = F\langle \xi, \eta \rangle$. We obviously have $\xi R \cap \eta R = 0$. Therefore R is not a right Ore domain (and neither a left Ore domain, i.e., the condition $Rr \cap Rs \neq 0$ for every $r, s \in R \setminus \{0\}$ is not fulfilled). $\qquad\qquad\square$

A domain that is neither right nor left Ore may still be embedded into a division ring, but not in such a way that condition (c) of Definition 7.10 would be fulfilled. The free algebra $F\langle \xi, \eta \rangle$ turns out to be an example of such a domain. On the other hand, there do exist domains that cannot be embedded into division rings. For details we refer the reader to [Lam99, Chap. 9].

The connection between $F\langle \xi, \eta \rangle$ and the nonfulfillment of the right Ore condition is even tighter.

Proposition 7.19 *Let A be a nonzero unital algebra over a field F. If A is a domain that is not right Ore, then A contains a subalgebra isomorphic to $F\langle \xi, \eta \rangle$.*

Proof By assumption there exist $r, s \in A \setminus \{0\}$ such that $rA \cap sA = 0$. Let A_0 be the unital subalgebra of A generated by r and s. We claim that the unital homomorphism $f(\xi, \eta) \mapsto f(r, s)$ from $F\langle \xi, \eta \rangle$ onto A_0 is injective, and hence an isomorphism. Suppose this is not true. Then there exists a nonzero polynomial $f(\xi, \eta)$ such that $f(r, s) = 0$. We may assume that f is of minimal degree. Let us write $f = \lambda + \xi g + \eta h$, where λ is the constant term of f and $g, h \in F\langle \xi, \eta \rangle$. There is no loss of generality in assuming that $h \neq 0$. Setting $u := g(r, s)$ and $v := h(r, s)$ we thus have $\lambda + ru + sv = 0$. Multiplying on the right by r we obtain $r(\lambda + ur) + svr = 0$, and so $r(\lambda + ur) = -svr \in rA \cap sA = 0$. Since $\deg(h) < \deg(f)$, $v \neq 0$. As s and r are also nonzero and A is a domain, we have arrived at a contradiction. $\qquad\square$

Trivial examples of right Ore domains are division rings and commutative domains. As another simple example we mention the ring of quaternions with integer coefficients (cf. Examples 7.9 and 7.11). The next proposition gives rise to less obvious examples.

Proposition 7.20 (Goldie) *A right noetherian domain R is a right Ore domain.*

Proof Take $r, s \in R \setminus \{0\}$, and set $J_n := rsR + r^2sR + \cdots + r^nsR$. Since the J_n's obviously form an ascending chain of right ideals, there exists $n \in \mathbb{N}$ such that $J_n = J_{n+1} = \ldots$. Pick any $x_0 \in R \setminus \{0\}$. Then $r^{n+1}sx_0$ is a nonzero element from J_{n+1}, and hence also from J_n. Thus there exists $k \leq n$ such that

$$r^k sx_k + r^{k+1}sx_{k+1} + \cdots + r^n sx_n = r^{n+1}sx_0$$

for some $x_k, \ldots, x_n \in R$ with $x_k \neq 0$. The cancellation property of R yields

$$sx_k + rsx_{k+1} + \cdots + r^{n-k}sx_n = r^{n+1-k}sx_0.$$

Therefore $sx_k \neq 0$ lies in rR. This shows that (7.2) holds.   □

Example 7.21 The Weyl algebra $\mathscr{A}_n$ is a noetherian domain (Example 6.4), therefore $Q_{rc}(\mathscr{A}_n)$ exists and is a division ring. Moreover, $Q_{rc}(\mathscr{A}_n) \cong Q_{lc}(\mathscr{A}_n)$ by Remark 7.16 and the left version of Proposition 7.20.

Proposition 7.20 is just a small extract from the theory developed by A. W. Goldie in the late 1950s. His main result characterizes rings whose right classical ring of quotients is isomorphic to $M_{n_1}(D_1) \times \cdots \times M_{n_r}(D_r)$ where D_i are division rings. A corollary of special importance is that right noetherian semiprime rings do have this property. Goldie's theory is treated in many books, including [Her68, Lam99, MR01, Row08].

7.4 Martindale Rings of Quotients

In this section we take another step forward in the study of rings of quotients—or perhaps a step backwards. Namely, the next construction is similar to, and in fact slightly simpler than the construction of $Q_{rc}(R)$. Roughly speaking, we will just replace the role of the right ideals containing regular elements by the (easier to handle) nonzero two-sided ideals. In spite of the technical similarities, the philosophy behind these two constructions is different. The goal of the next one is not making regular elements invertible, but merely providing a larger ring that can be used as a tool for solving problems concerning the original ring. Still, the field of quotients of a commutative domain serves as a prototype here as well. We will assume that our rings are prime. This is a natural setting; a commutative ring is a domain if and only if it is prime, and the role of prime rings among general rings is often parallel to the role of domains among commutative rings. With a little extra effort we could handle more general semiprime rings, but we will stick with the classical prime ring situation, as considered in the seminal work of W. S. Martindale from 1969.

Constructing the right Martindale ring of quotients $Q_r(R)$. Let $R \neq 0$ be a prime ring. The set $\mathscr{I}$ of all nonzero ideals of R is closed under products, and hence also under finite intersections. We endow the set of all pairs (f, I), where $I \in \mathscr{I}$ and $f : I \to R$ is a right R-module homomorphism, by the following relation, which is readily seen to be equivalence: $(f, I) \sim (g, J)$ if f and g coincide on some $K \in \mathscr{I}$ such that $K \subseteq I \cap J$. Write $[f, I]$ for the equivalence class determined by (f, I), and denote by $Q_r(R)$ the set of all equivalence classes, equipped with addition and multiplication

$$[f_1, I_1] + [f_2, I_2] := [f_1 + f_2, I_1 \cap I_2],$$
$$[f_1, I_1][f_2, I_2] := [f_1 f_2, I_2 I_1]$$

(note that f_1f_2 is indeed defined on I_2I_1 since $f_2(I_2I_1) = f_2(I_2)I_1 \subseteq I_1$). It is easy to verify that these operations are well-defined. Indeed, assume that $(f_1, I_1) \sim (g_1, J_1)$ and $(f_2, I_2) \sim (g_2, J_2)$, i.e., there exist $K_i \in \mathscr{J}$ such that $K_i \subseteq I_i \cap J_i$ and $f_i = g_i$ on $K_i, i = 1, 2$. Then $f_1 + f_2 = g_1 + g_2$ on $K_1 \cap K_2 \in \mathscr{J}$ and $f_1f_2 = g_1g_2$ on $K_2K_1 \in \mathscr{J}$. Thus, $(f_1 + f_2, I_1 \cap I_2) \sim (g_1 + g_2, J_1 \cap J_2)$ and $(f_1f_2, I_2I_1) \sim (g_1g_2, J_2J_1)$. One immediately checks that $Q_r(R)$ is a ring with zero element $[0, R]$ and unity $[\mathrm{id}_R, R]$.

Definition 7.22 The ring $Q_r(R)$ is called the **right Martindale ring of quotients** of R.

The **left** Martindale ring of quotients $Q_l(R)$ is constructed analogously through the left R-module homomorphisms. In general, $Q_r(R) \ncong Q_l(R)$.

Theorem 7.23 *Let* $R \neq 0$ *be a prime ring, and let* $\mathscr{J}$ *be the set of all nonzero ideals of* R. *The ring* $Q_r(R)$ *has the following properties:*

(a) *$Q_r(R)$ is a unital ring containing R as a subring.*
(b) *For every $q \in Q_r(R)$ there exists $I \in \mathscr{J}$ such that $qI \subseteq R$.*
(c) *For every $q \in Q_r(R)$ and $I \in \mathscr{J}$, $qI = 0$ implies $q = 0$.*
(d) *If $I \in \mathscr{J}$ and $f : I \to R$ is a right R-module homomorphism, then there exists $q \in Q_r(R)$ such that $f(x) = qx$ for all $x \in I$.*

Moreover, these properties characterize $Q_r(R)$ up to isomorphism.

Proof We already know that $Q_r(R)$ is a unital ring. Note that

$$\iota : R \to Q_r(R), \quad \iota(r) = [L_r, R],$$

where L_r is a left multiplication map, is an embedding of R into $Q_r(R)$ (injectivity follows from the primeness of R). Identifying R with its isomorphic copy $\iota(R)$ we may consider R as a subring of $Q_r(R)$, so that (a) holds.

Take $q = [f, I] \in Q_r(R)$. For every $x \in I$ we have

$$qx = [f, I][L_x, R] = [fL_x, RI] = [L_{f(x)}, R] = f(x) \in R.$$

This proves (b). Moreover, note that $qI = 0$ implies $f(I) = 0$ and hence $q = [f, I] = 0$. If I' is another member of $\mathscr{J}$ and $qI' = 0$, then also $(qI)I' \subseteq qI' = 0$. Since R is prime and $qI \subseteq R$, this yields $qI = 0$, and hence $q = 0$. Thus (c) is proved. Verifying (d) is also easy: If $I \in \mathscr{J}$ and $f : I \to R$ is a right R-module homomorphism, then, as noticed at the beginning of this paragraph, $q := [f, I] \in Q_r(R)$ satisfies $f(x) = qx$ for all $x \in I$.

Let Q be an arbitrary ring with properties (a)–(d). Take $q \in Q$. By (b) there exists $I \in \mathscr{J}$ such that $qI \subseteq R$. Therefore we can define the right R-module homomorphism $f : I \to R$ by $f(x) = qx$ for all $x \in I$. If $I' \in \mathscr{J}$ also satisfies $qI' \subseteq R$, then f coincides with $f' : I' \to R, f(x') = qx'$, on $I \cap I' \in \mathscr{J}$. Hence we see that the map

$$\varphi : Q \to Q_r(R), \quad \varphi(q) = [f, I],$$

is well-defined. It is easy to check that φ is a ring homomorphism. The injectivity of φ follows from (c), and the surjectivity from (d). Therefore $Q \cong Q_r(R)$. □

Further properties of $Q_r(R)$ can be extracted from the basic ones (a)–(d). For example, (c) can be extended as follows.

Lemma 7.24 *For every $q_1, q_2 \in Q_r(R)$ and $I \in \mathscr{J}$, $q_1 I q_2 = 0$ implies $q_1 = 0$ or $q_2 = 0$. Accordingly, $Q_r(R)$ is a prime ring.*

Proof By (b) we may choose $I_i \in \mathscr{J}$ such that $q_i I_i \subseteq R$, $i = 1, 2$. Hence we may conclude from $(q_1 I_1) I (q_2 I_2) \subseteq (q_1 I q_2) I_2 = 0$ that $q_1 I_1 = 0$ or $q_2 I_2 = 0$ (see Remark 2.25). From (c) we thus get $q_1 = 0$ or $q_2 = 0$. □

For further reference we record a straightforward extension of (b).

Lemma 7.25 *Given $q_1, \ldots, q_n \in Q_r(R)$ there exists $I \in \mathscr{J}$ such that $q_i I \subseteq R$, $i = 1 \ldots, n$.*

Proof By (b), for each i there exists $I_i \in \mathscr{J}$ such that $q_i I_i \subseteq R$. Hence $I := I_1 \cap \cdots \cap I_n$ has the desired property. □

Example 7.26 If R is a commutative domain, then, as already indicated, $Q_r(R)$ is isomorphic to the field of quotients $Q_Z(R)$ of R. We must show that $Q_Z(R)$ has properties (a)–(d). Only (d) is not entirely obvious. Assume, therefore, that $f : I \to R$ is a right R-module homomorphism. Take a nonzero $a \in I$. We have $f(x)a = f(xa) = f(ax) = f(a)x$ for every $x \in I$, and hence $f(x) = qx$ where $q := f(a)a^{-1} \in Q_Z(R)$.

Thus, for a commutative domain R we have $Q_Z(R) \cong Q_{cr}(R) \cong Q_r(R)$. The same is true if R is any of the rings from Examples 7.7–7.9, i.e., $M_n(\mathbb{Z})$, $\left[\begin{smallmatrix} \mathbb{Z} & 2\mathbb{Z} \\ \mathbb{Z} & \mathbb{Z} \end{smallmatrix}\right]$, or $\mathbb{Z} \oplus \mathbb{Z}i \oplus \mathbb{Z}j \oplus \mathbb{Z}k$. One can prove this from scratch. On the other hand, one can apply Corollary 7.61 below.

Example 7.27 If R is a simple unital ring, then $Q_r(R) = R$. Indeed, R obviously has properties (a)–(c), and checking that it also satisfies (d) is immediate: If $f : R \to R$ is a right R-module homomorphism, then $f(x) = f(1x) = f(1)x$ for all $x \in R$.

Example 7.28 Let V be an infinite dimensional vector space over a field F. Denote by K the subring of $\text{End}_F(V)$ consisting of all finite rank operators. Clearly, every operator in K is a sum of rank one operators $v \otimes h$, defined by $(v \otimes h)(u) = h(u)v$; here, $v \in V$ and $h \in V^*$, the dual of V (cf. Remark 4.17). It is an easy exercise to show that K is a simple ring without unity. Let R be any ring such that $K \subseteq R \subseteq \text{End}_F(V)$. Note that R is prime and that K is its minimal ideal. Let us show that

$$Q_r(R) \cong \text{End}_F(V).$$

We must verify that $\text{End}_F(V)$ has the four properties (a)–(d). The first three are easy: (a) is trivial, (b) follows from the fact that K is a (left) ideal of $\text{End}_F(V)$, and

(c) from the fact that every nonzero ideal of R contains K. To verify (d), take a nonzero ideal I of R and a right R-module homomorphism $f : I \to R$. Pick $v_0 \in V$ and $h_0 \in V^*$ so that $h_0(v_0) = 1$. Define $q : V \to V$ by $q(v) := f(v \otimes h_0)(v_0)$. Obviously, $q(v + v') = q(v) + q(v')$. Since

$$f(\lambda k)k' = f(\lambda k k') = f(k(\lambda k')) = \lambda f(k)k'$$

holds for all $\lambda \in F$ and $k, k' \in K$, i.e., $\big(f(\lambda k) - \lambda f(k)\big)K = 0$, it follows that $f(\lambda k) = \lambda f(k)$, $k \in K$. This yields $q(\lambda v) = \lambda q(v)$. That is, $q \in \mathrm{End}_F(V)$. Next, for all $v \in V$ and $h \in V^*$ we have

$$
\begin{aligned}
f(v \otimes h) &= f(v \otimes h_0 \cdot v_0 \otimes h) = f(v \otimes h_0) \cdot v_0 \otimes h \\
&= f(v \otimes h_0)(v_0) \otimes h = q(v) \otimes h \\
&= q \cdot v \otimes h.
\end{aligned}
$$

Thus, $f(k) = qk$ for every $k \in K$, and hence $f(x)k = f(xk) = qxk$ for all $x \in I$ and $k \in K$. That is, $(f(x) - qx)K = 0$, which yields $f(x) = qx$. This completes the proof.

7.5 The Extended Centroid

Throughout this section we assume that R is a nonzero prime ring.

Definition 7.29 The center of $Q_r(R)$ is called the **extended centroid** of R.

The extended centroid of R will be denoted by C. As usual, Z will stand for the center of R, and, as in the preceding section, $\mathscr{J}$ for the set of all nonzero ideals of R. We will continuously refer to properties (a)–(d) of Theorem 7.23.

Lemma 7.30 *If $q \in Q_r(R)$ is such that $qr = rq$ for all $r \in R$, then $q \in C$. Accordingly, Z is a subring of C.*

Proof Pick $q' \in Q_r(R)$. We must show that $[q, q'] = 0$. By (b) there exists $I \in \mathscr{J}$ such that $q'I \subseteq R$. Take $x \in I$. Since q commutes with x and $q'x$, we have $qq'x = q'xq = q'qx$. Thus, $[q, q']I = 0$, and hence $[q, q'] = 0$ by (c). $\square$

If elements in $Q_r(R)$ correspond to right R-module homomorphisms, then elements in C correspond to R-bimodule homomorphisms (these are, of course, maps that are both left and right R-module homomorphisms):

Lemma 7.31 *If $f : I \to R$, where $I \in \mathscr{J}$, is an R-bimodule homomorphism, then there exists $\lambda \in C$ such that $f(x) = \lambda x$ for all $x \in I$.*

Proof Since f is, in particular, a right R-module homomorphism, by (d) there exists $q \in Q_r(R)$ such that $f(x) = qx$, $x \in I$. On the other hand, since f is also a left R-module homomorphism it follows that $q(rx) = r(qx)$ for all $r \in R$ and $x \in I$. That is, $[q, r]I = 0$, and so $[q, r] = 0$ by (c). Therefore $\lambda := q \in C$ by Lemma 7.30. $\qquad\square$

Conversely, if $\lambda \in C$ and $I \in \mathscr{J}$ are such that $\lambda I \subseteq R$, then $x \mapsto \lambda x$ is an R-bimodule homomorphism from I into R. The extended centroid, although defined through the right Martindale ring of quotients, is thus a left-right symmetric notion.

Remark 7.32 Suppose R is an algebra over a field F. Every $\alpha \in F$ gives rise to the R-bimodule homomorphism $x \mapsto \alpha x$ from R into R. Using Lemma 7.31 one therefore easily infers that F embeds canonically in C. Accordingly, F can be considered as a subfield of C.

Theorem 7.33 *The extended centroid C of a nonzero prime ring R is a field.*

Proof Take $\lambda \neq 0$ in C and choose $I \in \mathscr{J}$ satisfying (b). Clearly, $\lambda I \in \mathscr{J}$ by (c). We claim that $f : \lambda I \to R, f(\lambda x) = x$, is well-defined. Indeed, $\lambda x = 0$ obviously implies $\lambda I_x = 0$, where I_x is the ideal of R generated by x, and hence $x = 0$ by (c). Since f is an R-bimodule homomorphism, Lemma 7.31 yields the existence of $\mu \in C$ such that $f(y) = \mu y, y \in \lambda I$. Writing $y = \lambda x, x \in I$, we thus get $\mu \lambda x = x$. That is, $(\mu\lambda - 1)I = 0$, and so $\mu\lambda = 1$ by (c). Thus, λ is invertible. $\qquad\square$

We may therefore regard $Q_r(R)$ as an algebra over C.

Definition 7.34 The C-subalgebra of $Q_r(R)$ generated by R is called the **central closure** of R.

The central closure of R, which will be denoted by R_C, obviously consists of elements of the form $\sum_i \lambda_i r_i$, where $\lambda_i \in C, r_i \in R$. Thus, R_C is, like C, a left-right symmetric object. The following symmetric version of Lemma 7.25 therefore holds for elements in R_C.

Lemma 7.35 *Given $q_1, \ldots, q_n \in R_C$, there exists $I \in \mathscr{J}$ such that $q_i I \subseteq R$ and $I q_i \subseteq R, i = 1 \ldots, n$.*

Proof It is enough to prove the lemma for $n = 1$. Namely, just as in the proof of Lemma 7.25, we can take the intersection of ideals corresponding to particular elements. Thus, pick $q = \sum_i \lambda_i r_i \in R_C$. By Lemma 7.25 there exists $I \in \mathscr{J}$ such that $\lambda_i I \subseteq R$ for each i. This clearly implies that $qI \subseteq R$ and, since λ_i commutes with elements from I, also $Iq \subseteq R$. $\qquad\square$

Assume that R has a nonzero center Z. Then the field of quotients $\widehat{Z}$ of Z may be viewed as a subfield of C. For clarity we record the following simple lemma.

Lemma 7.36 *If $Z \neq 0$, then $Q_Z(R)$ can be embedded in R_C. Moreover, if $\widehat{Z} = C$, then $Q_Z(R) \cong R_C$.*

Proof The first statement follows from Proposition 7.3, and the second one is obvious. $\qquad\square$

Rings from Example 7.26 satisfy the condition $\widehat{Z} = C$. In general, $\widehat{Z}$ is not equal to C, not even when Z is already a field and thus equal to $\widehat{Z}$.

Example 7.37 Let F_0 be a subfield of F. Consider the special case of Example 7.28 where $R = K + F_0 \cdot 1$ (here, 1 is the identity operator). Note that $Z = F_0 \cdot 1$, and hence $\widehat{Z} = Z \cong F_0$. In Example 7.28 we have shown that $Q_r(R) \cong \mathrm{End}_F(V)$, which implies that $C \cong F$. Accordingly, $R_C = K + F \cdot 1$.

This example indicates the intrinsic nature of the notion of the extended centroid. Namely, in the framework of the F-linear operators it is more natural to deal with F than with its subfields.

Definition 7.38 If $R = R_C$, then R is said to be **centrally closed**.

We remark that $Z \neq 0$ implies $C \subseteq R_C$. Indeed, $\lambda = (\lambda z^{-1})z \in R_C$ for every $\lambda \in C$, where $z \in Z \setminus \{0\}$. Therefore, in this case R is centrally closed if and only if $Z = C$. But R may be centrally closed also when $Z = 0$.

Example 7.39 Every simple ring R, regardless of whether its center is 0 or not, is centrally closed since $\lambda R \subseteq R$ for all $\lambda \in C$ by (b). A concrete example of a simple ring with zero center is the ring K of finite rank operators from Example 7.28.

By the term **"centrally closed prime algebra"** we will mean a centrally closed prime ring viewed as an algebra over its extended centroid. As pointed out in the preceding paragraph, the center of such an algebra A is either 0 or coincides with the extended centroid. In the latter case, A is a central algebra. The notion of a centrally closed prime algebra can be thus considered as a generalization of the notion of a central simple algebra. An example of a centrally closed prime algebra that is not simple is $\mathrm{End}_F(V)$ where V is an infinite dimensional vector space over F (cf. Example 7.28).

Lemma 7.40 *The central closure R_C of any prime ring R is a centrally closed prime algebra over C.*

Proof Since $R_C \supseteq R \in \mathscr{J}$, the primeness of R_C follows from Lemma 7.24. Let us denote the extended centroid of R_C by E. By Remark 7.32 we may consider C as a subfield of E. The lemma will be established by showing that $E = C$. Take $\varepsilon \in E$. By (b), there exists a nonzero ideal U of R_C such that $\varepsilon U \subseteq R_C$. Note that $I := \{x \in U \cap R \mid \varepsilon x \in R\}$ is an ideal of R. We claim that $I \neq 0$. Indeed, choose $u \neq 0$ in U and use Lemma 7.25 to get $J \in \mathscr{J}$ such that $uJ \subseteq R$ and $(\varepsilon u)J \subseteq R$. Then uJ is contained in I and is not 0 by (c). We can now define the R-bimodule homomorphism $f : I \to R, f(x) = \varepsilon x$. By Lemma 7.31 there exists $\lambda \in C$ such that $\varepsilon x = \lambda x, x \in I$. This readily implies that $\varepsilon y = \lambda y$ for every y from the subspace $\widetilde{I}$ of R_C generated by I. Observe that $\widetilde{I}$ is an ideal of R_C, so that (c) yields $\varepsilon = \lambda \in C$. $\square$

7.6 Linear (In)dependence in Prime Rings

Numerous applications of the Martindale rings of quotients stem from the results
we are about to establish. To get a feel for the topic, we first examine the condition
studied in the next lemma in the simple setting of division rings. Thus, let D be a
division ring, and let $a, b \in D$ be such that $axb = bxa$ for every $x \in D$. A natural
possibility where this occurs is when a and b are linearly dependent over the center
of D. Actually, this is the only possibility. Indeed, assuming that $a \neq 0$ and then
multiplying $axb = bxa$ by a^{-1} on both sides we get $xba^{-1} = a^{-1}bx$, which readily
implies that $\lambda := ba^{-1}(= a^{-1}b)$ lies in the center of D and satisfies $b = \lambda a$. Thus,
we have a simple connection between the multiplicative structure of D ($axb = bxa$)
and the linear structure of D ($b = \lambda a$). This little observation can be generalized in
various directions.

We use the same notation as in the preceding sections. In particular, R always
denotes a nonzero prime ring and $\mathscr{I}$ the set of all nonzero ideals of R. By (b) and
(c) we refer to conditions of Theorem 7.23.

Lemma 7.41 *Let* $a, b \in Q_r(R)$ *and let* $I \in \mathscr{I}$. *If* $axb = bxa$ *for all* $x \in I$, *then* a
and b *are linearly dependent over* C.

Proof By Lemma 7.25 there exists $J \in \mathscr{I}$ such that $aJ \subseteq R$ and $bJ \subseteq R$. By
replacing the role of I by $I \cap J$ we see that there is no loss of generality in assuming
that $aI \subseteq R$ and $bI \subseteq R$.

We may assume that $a \neq 0$. Then $aI \neq 0$ by (c), and hence $K := IaI \in \mathscr{I}$. We
claim that the map $f : K \to R$,

$$f\left(\sum_i x_i a y_i\right) = \sum_i x_i b y_i, \quad x_i, y_i \in I,$$

is well-defined. Assume that $\sum_i x_i a y_i = 0$. Multiplying on the right by $zb, z \in R$, and
using $a(y_i z)b = b(y_i z)a$, it follows that $\left(\sum_i x_i b y_i\right)za = 0$. But then $\sum_i x_i b y_i = 0$
since R is prime. Note also that f indeed maps into R for $bI \subseteq R$. Obviously, f is an
R-bimodule homomorphism, so it follows from Lemma 7.31 that there exists $\lambda \in C$
such that $f(w) = \lambda w$ for every $w \in K$. In particular, $xby = \lambda xay$ for all $x, y \in I$. That
is, $I(b - \lambda a)I = 0$, and hence $b = \lambda a$ by Lemma 7.24. $\square$

Even in the basic case where $a, b \in R$ and $I = R$, the involvement of the extended
centroid C in Lemma 7.41 is indispensable, i.e., it cannot be replaced by the center
Z even when Z is a field. Indeed, a glance at Example 7.37 convinces us of that.

The next lemma treats a considerably more general condition, but, as we shall
see, the proof can be easily reduced to the situation treated in Lemma 7.41.

Lemma 7.42 *Let* $a_i, b_i \in Q_r(R)$ *and* $I \in \mathscr{I}$ *be such that*

$$\sum_{i=1}^{n} a_i x b_i = 0 \text{ for all } x \in I.$$

If $a_1, \ldots, a_n$ are linearly independent over C, then each $b_i = 0$. Similarly, if $b_1, \ldots, b_n$ are linearly independent over C, then each $a_i = 0$.

Proof Let us only prove the first statement. The second statement can be handled similarly. Thus, assume that $a_1, \ldots, a_n$ are linearly independent. The $n = 1$ case follows from Lemma 7.24. We may therefore assume that the desired conclusion holds if the number of summands is smaller than n. Suppose that, say, $b_n \neq 0$. By (b) we can choose $J \in \mathscr{J}$ such that $b_n J \subseteq R$. Then $xb_n y \in I$ whenever $x \in I$ and $y \in J$, so that $\sum_{i=1}^n a_i (xb_n y) b_i = 0$. Since the last term, $a_n xb_n yb_n$, is equal to $-\left(\sum_{i=1}^{n-1} a_i xb_i\right) yb_n$, we can rewrite this identity as

$$\sum_{i=1}^{n-1} a_i x(b_n yb_i - b_i yb_n) = 0 \quad \text{for all } x \in I, y \in J.$$

The induction assumption yields $b_n yb_i - b_i yb_n = 0$ for all $y \in J$ and $i = 1, \ldots, n - 1$. As $b_n \neq 0$ by assumption, Lemma 7.41 tells us that there exist $\lambda_i \in C$ such that $b_i = \lambda_i b_n$, $i = 1, \ldots, n - 1$. But then we can rewrite $\sum_{i=1}^n a_i xb_i = 0$ as $\left(\sum_{i=1}^n \lambda_i a_i\right) xb_n = 0$, where $\lambda_n = 1$. Hence $\sum_{i=1}^n \lambda_i a_i = 0$ by Lemma 7.24, contradicting the linear independence of the a_i's. $\square$

Lemma 7.42 will play a crucial role in the rest of the chapter. Let us point out that this lemma generalizes Lemma 1.24, which was used as one of the main tools in the first chapters.

The following theorem is just a refined version of Lemma 7.42.

Theorem 7.43 *Let $a_i, b_i, c_j, d_j \in Q_r(R)$ and $I \in \mathscr{J}$ be such that*

$$\sum_{i=1}^n a_i xb_i = \sum_{j=1}^m c_j xd_j \quad \text{for all } x \in I.$$

If $a_1, \ldots, a_n$ are linearly independent over C, then each b_i is a C-linear combination of $d_1, \ldots, d_m$. Similarly, if $b_1, \ldots, b_n$ are linearly independent over C, then each a_i is a C-linear combination of $c_1, \ldots, c_m$.

Proof We only prove the first statement. The proof is practically the same as that of Lemma 4.9, but we give it anyway. Extend $\{a_1, \ldots, a_n\}$ to a basis of the C-linear span of $\{a_1, \ldots, a_n, c_1, \ldots, c_m\}$. Denote the additional basis elements (if there are any) by $a_{n+1}, \ldots, a_p$, and write each c_j as a C-linear combination of $a_1, \ldots, a_p$. Our identity can then be rewritten as $\sum_{i=1}^p a_i xb_i' = 0$, where b_i', $i \leq n$, is the sum of b_i and a linear combination of the d_j's. Lemma 7.42 shows that each $b_i' = 0$. $\square$

Mentioning a result on tensor products (Lemma 4.9) in the last proof was no coincidence. The above results can be interpreted in terms of tensor products and multiplication algebras. The next theorem is a generalization of Theorem 4.26 which considers central simple algebras.

Theorem 7.44 *If A is a centrally closed prime algebra, then $A \otimes A^\circ \cong M(A)$ under the map $a \otimes b \mapsto L_a R_b$.*

Proof The proof is literally the same as that of Theorem 4.26. The only difference is that one has to refer to Lemma 7.42 instead of to Lemma 1.24. $\qquad\square$

The next theorem and its corollary make a passage to the topics of the next sections. The theorem shows that the linear dependence of elements in $Q_r(R)$ can be characterized through an identity involving the Capelli polynomials c_m (introduced in Sect. 6.3).

Theorem 7.45 *Elements $a_1, \ldots, a_m \in Q_r(R)$, $m \geq 2$, are linearly dependent over C if and only if $c_m(a_1, \ldots, a_m, x_1, \ldots, x_{m-1}) = 0$ for all $x_1, \ldots, x_{m-1} \in R$.*

Proof Since c_m is alternating in the first m indeterminates, the "only if" part follows from Lemma 6.9. The "if" part will be proved by induction on m. The $m = 2$ case is exactly the content of Lemma 7.41 (with $I = R$). We may thus assume that there exist $x_2, \ldots, x_{m-1} \in R$ such that $c_{m-1}(a_1, \ldots, a_{m-1}, x_2, \ldots, x_{m-1}) \neq 0$. Applying the formula (6.2) from Sect. 6.3 we see that $c_m(a_1, \ldots, a_m, x_1, \ldots, x_{m-1}) = 0$ can be written as

$$\sum_{i=1}^{m} (-1)^{i-1} a_i x_1 c_{m-1}(a_1, \ldots, a_{i-1}, a_{i+1}, \ldots, a_m, x_2, \ldots, x_{m-1}) = 0,$$

and so the linear dependence of $a_1, \ldots, a_m$ follows from Lemma 7.42. $\qquad\square$

Note that the theorem remains valid, with the same proof, if the x_i's are taken from a nonzero ideal of R.

Corollary 7.46 *For every $n \geq 2$, the Capelli polynomial c_{n^2} is an identity of every proper subalgebra of $M_n(F)$, but not of $M_n(F)$ itself.*

Proof A proper subalgebra B of $M_n(F)$ has dimension less than n^2, so c_{n^2} is an identity of B by the "only if" part of Theorem 7.45 (the extended centroid of $M_n(F)$ is of course F, cf. Example 7.27). Conversely, choosing linearly independent $a_1, \ldots, a_{n^2} \in M_n(F)$ it follows by the "if" part that there exist $x_i \in M_n(F)$ such that $c_{n^2}(a_1, \ldots, a_{n^2}, x_1, \ldots, x_{n^2-1}) \neq 0$. $\qquad\square$

7.7 Prime GPI-Rings

We now change the subject of discussion. The theory developed in the previous sections will still be present, but just as a tool used in the proofs. The central topic from now on will be rings satisfying polynomial identities and generalized polynomial identities. Although it may appear strange at a glance, we will first treat the latter and after that the former. This order is more suitable for our approach.

Informally, one can think of a **generalized polynomial identity** as of a polynomial identity in which some of the indeterminates are replaced by fixed elements. Thus, there exist elements a_{i_k} such that

$$\sum a_{i_0} x_{j_1} a_{i_1} \ldots a_{i_{n-1}} x_{j_n} a_{i_n} = 0 \qquad (7.3)$$

holds for arbitrary x_{j_l}.

Example 7.47 Let R be the ring from Example 7.28, i.e., R is any ring such that $K \subseteq R \subseteq \operatorname{End}_F(V)$ where V is an infinite dimensional vector space over F and K is the subring of $\operatorname{End}_F(V)$ consisting of all finite rank operators. In fact, what we are about to say also makes sense if V is finite dimensional (in this case we clearly have $R = \operatorname{End}_F(V) \cong M_n(F)$), we are now just more interested in rings that do not satisfy nontrivial polynomial identities. Take any rank one operator $a \in K$. One can readily check that for every $x \in R$ there is $\lambda_x \in F$ such that $axa = \lambda_x a$. Given $x, y \in R$, we thus have $(axa)ya = \lambda_x \lambda_y a = ay(axa)$. Thus,

$$axaya = ayaxa \quad \text{for all } x, y \in R. \qquad (7.4)$$

This is a model of a generalized polynomial identity.

We will confine ourselves to prime rings. Otherwise $axb = 0$, $x \in R$, was a nontrivial generalized polynomial identity, which would make the subject rather muddled. The standard definition of a prime GPI-ring requires the introduction of an appropriate generalization of a free algebra, which provides a suitable setting for "generalized polynomials". To make our exposition as simple as possible, we will avoid this and give a rather straightforward definition which involves only multilinear identities, i.e., such that each x_j appears in each term of (7.3) exactly once (as for example in (7.4)). More precisely, by a **multilinear generalized polynomial identity** of a prime ring R we mean an identity of the form

$$\sum_{\sigma \in S_n} \sum_{i=1}^{n_\sigma} a_{0i}^\sigma x_{\sigma(1)} a_{1i}^\sigma x_{\sigma(2)} a_{2i}^\sigma \ldots a_{n-1,i}^\sigma x_{\sigma(n)} a_{ni}^\sigma = 0 \qquad (7.5)$$

for all $x_1, \ldots, x_n \in R$, where a_{ti}^σ are fixed elements in $Q_r(R)$. Assuming that the a_{ti}^σ's belong to R does not make the treatment much easier, and because of applications it is really useful to involve $Q_r(R)$. Applying the linearization process, which works just as well as for ordinary polynomial identities, it is possible to show that restricting to multilinear identities does not cause loss of generality; our definition of a prime GPI-ring will in fact be equivalent to the standard one (as given in the treatise [BMM96] of the theory of generalized identities). Before giving it, let us ask ourselves what should be required in order to consider the identity (7.5) as nontrivial, in the sense that it can yield some information about the structure of R. Assuming that every $a_{ti}^\sigma \neq 0$ is certainly not enough. Say, $ax - xa = 0$ only means that a is in the center of R, and says nothing about R. More generally, from Lemma 7.42 (and Theorem 7.43) we know that the identities of the form $\sum_i a_i x b_i = 0$ are merely consequences of the linear dependence relations among the a_i's and the b_i's. A similar statement can be made for any multilinear generalized polynomial identity in which the x_i's appear in the same order in each of the terms, i.e., $\sum_i a_{0i} x_1 a_{1i} x_2 a_{2i} \ldots a_{n-1,i} x_n a_{ni} = 0$ (cf. Theorem 7.45 and its proof). Therefore, the least one has to require in order to regard (7.5) as nontrivial is that one of the summations

$$\sum_{i=1}^{n_\sigma} a_{0i}^\sigma x_{\sigma(1)} a_{1i}^\sigma x_{\sigma(2)} a_{2i}^\sigma \ldots a_{n-1,i}^\sigma x_{\sigma(n)} a_{ni}^\sigma \qquad (7.6)$$

is not always zero (i.e., is not an identity itself). As we shall see, this is already suffi-cient. Let us call (7.6) the σ-**term** of (7.5). Every multilinear generalized polynomial identity is thus the sum of its σ-terms.

Definition 7.48 A nonzero prime ring R is said to be a **GPI-ring** if there exists a multilinear generalized polynomial identity of R such that one of its σ-terms is not an identity of R. For convenience, we also consider $R = 0$ as a GPI-ring.

The ring R from Example 7.47 is thus a GPI-ring since $axaya \neq 0$ for some $x, y \in R$. We remark that the center of R may be 0 (say, if $R = K$), and that the extended centroid of R may properly contain the field of quotients of the center of R (Example 7.37). This shows that prime GPI-rings do not share all the nice properties of prime PI-rings that will be studied in the next sections; cf. Theorem 7.56 and Corollary 7.57.

The study of generalized polynomial identities was initiated by S. A. Amitsur. In the first paper on the subject, in 1965, he described primitive GPI-rings. Our goal in this section is to prove a more general result on prime GPI-rings, established by W. S. Martindale in 1969.

As always, C stands for the extended centroid of a prime ring R, R_C for the central closure of R, and $Q_r(R)$ for the right Martindale ring of quotients of R.

Lemma 7.49 *If R is a nonzero prime GPI-ring, then there exist $a, b \in R \setminus \{0\}$ such that $\dim_C aR_C b < \infty$.*

Proof Among multilinear generalized polynomial identities of R whose σ-terms are not all identities, choose one with the minimal number of summands. There must be at least two permutations, σ_1 and σ_2, such that the corresponding σ_1-term and σ_2-term are not identities. As $\sigma_1^{-1} \neq \sigma_2^{-1}$, we may assume without loss of generality that $\sigma_1^{-1}(1) < \sigma_1^{-1}(2)$ and $\sigma_2^{-1}(2) < \sigma_2^{-1}(1)$. As usual, let $x_1, \ldots, x_n$ be variables in our identity. Denote by $f_1(x_1, \ldots, x_n)$ the sum of all σ-terms such that $\sigma^{-1}(1) < \sigma^{-1}(2)$, and by $f_2(x_1, \ldots, x_n)$ the sum of all σ-terms such that $\sigma^{-1}(2) < \sigma^{-1}(1)$. Thus, $f_1(x_1, \ldots, x_n)$ consists of summands in which x_1 appears before x_2, and $f_2(x_1, \ldots, x_n)$ consists of summands in which x_2 appears before x_1. Of course, $f_1(x_1, \ldots, x_n) + f_2(x_1, \ldots, x_n) = 0$ for all $x_1, \ldots, x_n \in R$. In view of the minimality assumption there exist $w_1, \ldots, w_n \in R$ such that $f_1(w_1, \ldots, w_n) \neq 0$. Note that the identity $f_1(x, y, w_3, \ldots, w_n) = -f_2(x, y, w_3, \ldots, w_n)$, where x, y are arbitrary and the w_i's are fixed, can be written as

$$\sum_{i=1}^{l} p_i x q_i y r_i = \sum_{j=1}^{m} s_j y t_j x u_j \quad \text{for all } x, y \in R \tag{7.7}$$

and some $p_i, q_i, r_i, s_j, t_j, u_j \in Q_r(R)$. We may assume that $\{p_1, \ldots, p_k\}$ is a maximal linearly independent subset of $\{p_1, \ldots, p_l\}$. Expressing each p_i, $i > k$, as a linear combination of $p_1, \ldots, p_k$, we see that (7.7) gets the form

$$\sum_{i=1}^{k} p_i x h_i(y) = \sum_{j=1}^{m} s_j y t_j x u_j \quad \text{for all } x, y \in R, \tag{7.8}$$

where h_i belong to $M(Q_r(R))$, the multiplication algebra of $Q_r(R)$. We are now in a position to apply Theorem 7.43. Accordingly, each $h_i(y)$ is a C-linear combination of $u_1, \ldots, u_m$. This holds for every $y \in R$, but since the elements in $A := R_C$ are just C-linear combinations of elements in R, it actually holds for every $y \in A$. As the summations in (7.8) are not 0 if we take $x = w_1$ and $y = w_2$, $h_i(y) \neq 0$ for some i and y. To summarize, the algebra $M(Q_r(R))$ contains nonzero elements that map A into some finite dimensional spaces. Among them choose h so that it can be written as $h = \sum_{i=1}^{d} L_{a_i} R_{b_i}$ with d minimal. We have to prove that $d = 1$. Suppose this is not true, i.e., $d > 1$. We will derive a contradiction by a similar trick as in the proof of Lemma 7.42. First observe that the minimality of d implies that the sets $\{a_1, \ldots, a_d\}$ and $\{b_1, \ldots, b_d\}$ are linearly independent. By Lemma 7.41 there exists $r \in R$ such that $b_d r b_1 - b_1 r b_d \neq 0$. Note that $h' := h R_{b_d r} - R_{r b_d} h \in M(Q_r(R))$ can be written as $h' = \sum_{i=1}^{d-1} L_{a_i} R_{b_i'}$ where $b_i' = b_d r b_i - b_i r b_d$. But this contradicts the assumption that d is minimal. Namely, $\dim_C h(A) < \infty$ implies $\dim_C h'(A) < \infty$, and the linear independence of $a_1, \ldots, a_{d-1}$ together with $b_1' \neq 0$ implies $h' \neq 0$ by Lemma 7.42.

Thus, there exist nonzero $a_1, b_1 \in Q_r(R)$ such that $\dim_C a_1 A b_1 < \infty$. This readily implies that $\dim_C a_1 r A b_1 s < \infty$ for any $r, s \in R$. By Theorem 7.23 (b)–(c), we may choose r and s so that $a := a_1 r$ and $b := b_1 s$ are nonzero elements in R. $\quad\square$

We are now just a step away from Martindale's theorem. However, let us pause for a moment and examine the special case where R is a simple unital ring. This is needed for applications to PI-rings in the next section.

Lemma 7.50 *A simple unital GPI-ring R is a finite dimensional algebra over its center Z.*

Proof Recall that $Q_r(R) = R$ and hence $C = Z$, the center of R (cf. Example 7.27). By Lemma 7.49 there exist $a, b \in R \setminus \{0\}$ such that $L_a R_b$ is a finite rank operator. Since R is simple and unital, $\sum_j u_j a v_j = \sum_k w_k b z_k = 1$ for some $u_j, v_j, w_k, z_k \in R$. Consequently, $\sum_{j,k} L_{u_j} R_{z_k} (L_a R_b) L_{v_j} R_{w_k}$ is the identity operator, and has finite rank. This means that $[R : Z] < \infty$. $\quad\square$

Theorem 7.51 (Martindale) *A nonzero prime ring R is a GPI-ring if and only if $A := R_C$ is a primitive ring containing an idempotent e such that Ae is a minimal left ideal of A and eAe is a finite dimensional division algebra over C.*

Proof We start with the easier "if" part. Let $d := [eAe : C]$. Then

$$s_{d+1}(ex_1e, \ldots, ex_{d+1}e) = 0$$

for all $x_1, \ldots, x_{d+1} \in R$ (see Example 6.15). This is a generalized polynomial identity whose 1-term $ex_1ex_2e \ldots ex_ne$ is not identically zero since R is prime.

We proceed to the "only if" part. By Lemma 7.49, there exist $a, b \in R \setminus \{0\}$ such that $\dim_C aAb < \infty$. If $L \neq 0$ is a left ideal of A such that $L \subseteq Ab$ and $M \neq 0$ is a right ideal of A such that $M \subseteq aA$, then $ML \subseteq aAb$ and hence $\dim_C ML < \infty$; moreover, $ML \neq 0$ for A is prime (Lemma 7.40). Choose L and M so that ML has minimal dimension. We claim that AML is a minimal left ideal of A. Let L' be a left ideal such that $0 \neq L' \subseteq AML$. Then $L' \subseteq L$; hence $ML' \subseteq ML$, and so $ML' = ML$ by the dimension assumption. Consequently, $L' \supseteq AML' = AML$, proving that AML is indeed minimal. We can now invoke Lemma 2.58 to obtain an idempotent $e \in A$ such that $Ae = AML$ and eAe is a division algebra. Moreover, using $e \in AML \subseteq AaAb$ it can be easily shown that eAe is finite dimensional over C. Finally, Lemma 5.5 tells us that A is a primitive ring. □

The results from Sect. 5.4 therefore hold for the central closure R_C of a prime GPI-ring R. Moreover, the associated division ring of R_C is a finite dimensional algebra in this context. A general prime GPI-ring is therefore quite close to the one considered in Example 7.47.

7.8 Primitive PI-Rings

We proceed with polynomial identities. So far we have considered them only in algebras, not yet in rings. Thus, the first question that presents itself is how to define a polynomial identity of a ring. An obvious choice is to give the same definition as for algebras, but requiring that the noncommutative polynomial in question has coefficients in $\mathbb{Z}$ (instead of in F). This is all right, but the problem, then, is identifying identities that should be treated as nontrivial. For instance, if a ring R has characteristic p, then the polynomial $p\xi$ vanishes on R. As there is no reason why an algebra of characteristic p should be a PI-algebra, we cannot regard $p\xi$, in spite of being a nonzero polynomial in $\mathbb{Z}\langle\xi\rangle$, as a nontrivial identity of R. Any definition of a nontrivial polynomial identity of a ring is therefore somewhat technical. One possibility is to admit those polynomials whose coefficient at one of their monomials of highest degree is 1. Assuming that such a polynomial is an identity and then employing the linearization process, one derives a multilinear identity in $\mathbb{Z}\langle\xi_1, \xi_2, \ldots\rangle$ such that one of its coefficients is 1. This can be easily shown by inspecting the proof of

Theorem 6.24, but there is no absolute need for us to bother with details. Let us simply restrict our attention to multilinear identities. Unlike in Chap. 6, we are now interested in the structure of rings satisfying some nontrivial polynomial identity, rather than in particular polynomial identities. Therefore we can afford this restriction without any loss of generality.

Definition 7.52 A ring R is said to be a **PI-ring** if there exists

$$f = \sum_{\sigma \in S_n} k_\sigma \xi_{\sigma(1)} \cdots \xi_{\sigma(n)} \in \mathbb{Z}\langle \xi_1, \xi_2, \ldots \rangle$$

such that $k_1 = 1$ and f is a **(polynomial) identity** of R, i.e., $f(x_1, \ldots, x_n) = 0$ for all $x_1, \ldots, x_n \in R$.

Of course, it is only important that one of the k_σ's equals 1. But it is convenient to require that $k_1 = 1$. This does not harm the generality for we can always rename the indeterminates.

If R is an algebra over a field, then R is a PI-algebra in the sense of Definition 6.13 if and only if R is a PI-ring in the sense of Definition 7.52. This is not obvious, but we omit the proof.

We will consider only *prime PI-rings*. This will make it possible for us to develop a decent structure theory. Without any restriction on rings we would face examples such as nilpotent rings, Boolean rings, the Grassmann algebra, etc. This is just too much to handle.

We will study general prime PI-rings in the next section, while this one is devoted to a smaller class of primitive PI-rings. What are the examples of such rings? Recall that there are two basic examples of PI-algebras: commutative algebras and finite dimensional algebras. Since a commutative primitive ring is a field (Lemma 5.7), and hence a 1-dimensional algebra over itself, the only immediate examples of primitive PI-rings are finite dimensional primitive (and hence simple, see Corollary 5.19) algebras. We will show that these obvious examples are, in fact, the only examples. For simple unital rings, which form a subclass of primitive rings (see Lemma 5.6), this follows immediately from Lemma 7.50. Namely, every prime PI-ring R is also a GPI-ring. Indeed, the identity $\sum_{\sigma \in S_n} k_\sigma x_{\sigma(1)} \cdots x_{\sigma(n)} = 0$, $k_1 = 1$, can be regarded as a multilinear generalized polynomial identity (7.5) with a_{ti}^σ being integer multiples of 1; if $R \neq 0$, then its 1-term $x_1 \ldots x_n$ is not an identity since R, as a prime ring, is not nilpotent. Thus:

Lemma 7.53 *A simple unital PI-ring R is a finite dimensional algebra over its center Z.*

The assumption that R is unital is actually redundant. Theorem 7.56 below shows that a simple PI-ring has a nonzero center, from which one easily deduces that it is automatically unital. Anyway, Lemma 7.53 in its present form is sufficient for our purposes. We will use it not only in the proof of the next theorem, but also in the next section. Since this lemma was (indirectly) derived from Lemma 7.49 which concerns prime GPI-rings, one might wonder whether the proof of the latter becomes simpler if we adapt it to the case where R is a simple unital PI-ring. However, besides notational simplifications and the omission of the last paragraph, there are no essential differences.

The following theorem was proved in 1948 by I. Kaplansky in his pioneering paper on polynomial identities.

Theorem 7.54 (Kaplansky) *A primitive PI-ring R is a finite dimensional simple algebra over its center Z.*

Proof In light of the Jacobson Density Theorem (Theorem 5.16), we may assume that R is a dense ring of linear operators of a vector space V over a division ring Δ. We will now mimic the proof of Lemma 6.38. Let f be a polynomial as in Definition 7.52. Suppose there exist linearly independent vectors $u_0, u_1, \ldots, u_n \in V$. By density, we can choose $h_1, \ldots, h_n \in R$ so that $h_i(u_j) = \delta_{ij} u_{j-1}$, $1 \leq i, j \leq n$. Then $h_1 \ldots h_n(u_n) = u_0$, while for any product of the h_i's in a different order we have $h_{i_1} h_{i_2} \ldots h_{i_n}(u_n) = 0$. Hence $f(h_1, \ldots, h_n)(u_n) = u_0 \neq 0$, a contradiction. Consequently, $\dim_\Delta V \leq n$, and so $R = \mathrm{End}_\Delta(V)$ (Corollary 5.13). Thus, R is a simple unital ring (cf. Theorem 3.31 and Example 1.10). Now apply Lemma 7.53. $\qquad\square$

7.9 Prime PI-Rings

The class of prime PI-rings is substantially larger than the class of primitive PI-rings. First of all, it contains all commutative domains. Rings from Examples 7.7–7.9, i.e., $M_n(\mathbb{Z})$, $\left[\begin{smallmatrix} \mathbb{Z} & 2\mathbb{Z} \\ \mathbb{Z} & \mathbb{Z} \end{smallmatrix}\right]$, and $\mathbb{Z} \oplus \mathbb{Z}i \oplus \mathbb{Z}j \oplus \mathbb{Z}k$ are all examples of prime PI-rings. Indeed, proving that they are prime is easy, and, as subrings of finite dimensional algebras ($M_n(\mathbb{Q})$, $M_2(\mathbb{Q})$, and $\mathbb{H}$, respectively) which satisfy standard polynomials, they are PI-rings. A description of prime PI-rings is therefore necessarily somewhat entangled, especially when compared with that of primitive PI-rings. Examples just given indicate that rings of quotients naturally come into play in this context.

Our first lemma already says a lot, yet it does not give the full picture. We use the standard notation: R is a nonzero prime ring with center Z, extended centroid C, and central closure R_C.

Lemma 7.55 *If R is a prime PI-ring, then $A := R_C$ is a finite dimensional central simple algebra.*

Proof Let U be a nonzero ideal of A. Every multilinear identity of R is obviously also an identity of A, and hence also of U. Let $f = \sum_{\sigma \in S_n} k_\sigma \xi_{\sigma(1)} \cdots \xi_{\sigma(n)} \in \mathbb{Z}\langle \xi_1, \xi_2, \ldots \rangle$, $k_1 = 1$, be a multilinear identity of U of minimal degree n.

Write

$$f = g\xi_n + \sum_i g_i\xi_n m_i, \tag{7.9}$$

where g, g_i are multilinear polynomials and each m_i is a monomial with leading coefficient 1 and of degree at least 1. Since g has degree $n - 1$ and $\xi_1 \ldots \xi_{n-1}$ is one of its monomials, g is not an identity of U. Let $u_1, \ldots, u_{n-1} \in U$ be such that $u := g(u_1, \ldots, u_{n-1}) \neq 0$. According to (7.9), $f(u_1, \ldots, u_{n-1}, x) = 0$ can be written as $ux = \sum v_i x w_i$ for all $x \in U$ and some $v_i \in A \cup \mathbb{Z}$, $w_i \in U$. This in particular holds for all x in $R \cap U$, which is an ideal of R different from 0 (the latter follows from Theorem 7.23). Writing ux as $ux1$, we thus infer from Theorem 7.43 that 1 is a C-linear combination of the w_i's. Therefore $1 \in A$ and so $C \subseteq A$, yielding $1 \in \sum_i Cw_i \subseteq U$. Thus $U = A$, showing that A is a simple unital algebra. By Lemma 7.53, A is finite dimensional over its center (which is equal to C by Lemma 7.30).   □

Given an arbitrary ring R and its proper ideal I, it is quite likely that I does not contain nonzero elements from the center Z of R. For example, this is certainly the case if Z is a field. On the other hand, take $M_n(\mathbb{Z})$, which is a typical example of a prime PI-ring. One can easily check that its nonzero ideals have the form $M_n(k\mathbb{Z})$ for some $k \in \mathbb{N}$ (cf. Corollary 4.43). Therefore each of them contains nonzero integer multiples of the identity matrix. Thus, nonzero ideals of $M_n(\mathbb{Z})$ intersect the center of $M_n(\mathbb{Z})$ nontrivially. This is actually true for every prime PI-ring:

Theorem 7.56 *If* $I \neq 0$ *is an ideal of a prime PI-ring* R, *then* $I \cap Z \neq 0$.

Proof Take an arbitrary nonzero C-linear functional γ on $A := R_C$. Since we know by Lemma 7.55 that A is a finite dimensional central simple algebra, Lemma 1.25 shows that there exists $g \in M(A)$ such that $g(x) = \gamma(x)1$, $x \in A$. Choose $a_i, b_i \in A$ so that $g = \sum_{i=1}^n L_{a_i}R_{b_i}$ and the sets $\{a_1, \ldots, a_n\}$ and $\{b_1, \ldots, b_n\}$ are linearly independent (cf. Remark 1.23). Lemma 7.35 implies that there exists a nonzero ideal J of R such that, in particular, $a_iJ \subseteq R$ and $Jb_i \subseteq R$ for every i. Then $K := JIJ$ is also a nonzero ideal of R and satisfies $g(K) \subseteq I \cap C$. Since $g(K) \neq 0$ by Lemma 7.42, it follows that $I \cap C \neq 0$. Of course, $I \cap C = I \cap Z$ for $I \subseteq R$.   □

Theorem 7.56 actually holds even for semiprime rings. This was proved by L. H. Rowen in 1973 (for the proof see, e.g., [Row08, p. 419]).

Corollary 7.57 *If* R *is a prime PI-ring, then* $Z \neq 0$ *and* $C = \widehat{Z}$, *the field of quotients of* Z. *Moreover, if* Z *is a field, then* R *is a finite dimensional central simple algebra.*

Proof Taking R for I in Theorem 7.56 we obtain $Z \neq 0$. Let $\lambda \in C$. By (b) (of Theorem 7.23) there exists an ideal $I \neq 0$ of R such that $\lambda I \subseteq R$. Using Theorem 7.56 again we obtain a nonzero $z \in I \cap Z$. Hence $\lambda z \in R \cap C = Z$, and so $\lambda = z'z^{-1}$ with $z, z' \in Z$. This proves that $C = \widehat{Z}$. Finally, if Z is a field, then $Z = \widehat{Z} = C$, and hence $R = R_C$ is a finite dimensional central simple algebra by Lemma 7.55.   □

The next theorem sharpens Lemma 7.55. It is usually attributed to E. Posner who proved a slightly weaker version in 1960.

Theorem 7.58 (Posner) *If R is a prime PI-ring, then $Q_Z(R)$ is a finite dimensional central simple algebra.*

Proof Corollary 7.57, together with Lemma 7.36, shows that $Q_Z(R) \cong R_C$. Now use Lemma 7.55. $\qquad\qquad\square$

By Wedderburn's theorem, $Q_Z(R)$ is isomorphic $M_n(D)$ for some finite dimensional central division algebra D and $n \in \mathbb{N}$. The point of Theorem 7.58 is that every prime PI-ring is very close to such an algebra. Namely, every element in $Q_Z(R)$ can be written as rz^{-1} where $r \in R$ and $z \in Z$. For example, every matrix in $M_n(\mathbb{Q})$ is of the form $\frac{1}{z}T$, where $z \in \mathbb{Z} \setminus \{0\}$ and $T \in M_n(\mathbb{Z})$.

The converse to Theorem 7.58 is trivially true: If $Q_Z(R)$ is finite dimensional, then it satisfies a standard identity so that it is a PI-ring, and so is its subring R. A somewhat rougher characterization of prime PI-rings reads as follows.

Corollary 7.59 *A prime ring R is a PI-ring if and only if R can be embedded into $M_m(K)$ for some $m \in \mathbb{N}$ and some field K.*

Proof The "if" part is clear. Let R be a PI-ring. Then $Q_Z(R)$ is a finite dimensional simple algebra over its center $\widehat{Z}$. Let K be a splitting field for $Q_Z(R)$ (for example, we can take an algebraic closure of $\widehat{Z}$). The scalar extension of $Q_Z(R)$ to K is then isomorphic to $M_m(K)$. $\qquad\qquad\square$

Remark 7.60 If R is a prime PI-algebra over a field F, then F is a subfield of $\widehat{Z}$ (see Remark 7.32), and so K is an extension field of F.

Corollary 7.59 implies that every prime PI-ring R satisfies a standard polynomial. In fact, if R embeds in $M_m(K)$, then, by the Amitsur-Levitzki Theorem, R satisfies s_{2m}.

As observed in the proof of Theorem 7.58, $Q_Z(R) \cong R_C$ holds for a prime PI-ring. For clarification, we add the following corollary.

Corollary 7.61 *If R is a prime PI-ring, then $Q_Z(R) \cong Q_r(R) \cong Q_{cr}(R)$.*

Proof It is clear that $Q_Z(R)$ satisfies conditions (a)–(c) of Theorem 7.23. It remains to prove (d). Take a nonzero ideal I of R and a right R-module homomorphism $f : I \to R$. By Theorem 7.56 there exists $0 \neq c \in I \cap Z$. We have $f(c)x = f(cx) = f(xc) = f(x)c$ for every $x \in I$, and so $f(x) = c^{-1}f(c)x$. Since $c^{-1}f(c) \in Q_Z(R)$, this proves that $Q_Z(R) \cong Q_r(R)$.

To prove that $Q_{cr}(R) \cong Q_Z(R)$, we have to verify that $Q_Z(R)$ satisfies conditions of Definition 7.10. Only the fulfillment of (b) is not entirely obvious. Take a regular element s in R. Then s is clearly also regular in $Q_Z(R)$. But then, since $Q_Z(R)$ is a finite dimensional unital algebra, s is invertible in $Q_Z(R)$ (cf. Remark 1.31 (b)). This proves (b). $\qquad\qquad\square$

7.10 Central Polynomials

In this final section we apply the above theory to a different theme.

Definition 7.62 Let A be an algebra over a field F. A polynomial $f(\xi_1, \ldots, \xi_n) \in F\langle \xi_1, \xi_2, \ldots \rangle$ with zero constant term is said to be a **central polynomial** for A if $f(x_1, \ldots, x_n)$ lies in the center of A for all $x_1, \ldots, x_n \in A$, but f is not an identity of A.

In other words, $f(\xi_1, \ldots, \xi_n)$ is a central polynomial if f is not an identity, but $[f, \xi_{n+1}]$ is. This shows that central polynomials can exist only on PI-algebras. The condition that f must have zero constant term excludes the trivial possibility where f is a sum of a constant polynomial and an identity.

Example 7.63 The polynomial $[\xi_1, \xi_2]$ is a central polynomial for the Grassmann algebra G; see Example 6.16.

Example 7.64 The polynomial $[\xi_1, \xi_2]^2$ is a central polynomial for the algebra $M_2(F)$. One can check this by a short calculation. On the other hand, this follows from the Cayley-Hamilton Theorem applied to the commutator of two 2×2 matrices. Indeed, recall that the characteristic polynomial is given by the formula (6.10) from Sect. 6.8, and that the trace of a commutator is 0. A similar, but multilinear polynomial $[\xi_1, \xi_2][\xi_3, \xi_4] + [\xi_3, \xi_4][\xi_1, \xi_2]$ is also central for $M_2(F)$.

There is no obvious way to generalize Example 7.64 to matrices of larger size. In fact, finding examples of central polynomials for $M_n(F)$ is not an easy task. However, they do exist, as we will soon find out.

Let us recall a few facts about the T-ideal $I_n := \mathrm{Id}(M_n(F))$ from Sect. 6.7. We assume, until further notice, that F is an infinite field. The relatively free algebra $F\langle X \rangle / I_n$ is then isomorphic to the algebra of $n \times n$ generic matrices $\mathrm{GM}_n(F)$; this is the subalgebra of $M_n(F[\Omega])$, $\Omega = \{\Omega_{jk}^{(i)} \mid j, k = 1, \ldots, n, \ i = 1, 2, \ldots\}$, generated by all $n \times n$ generic matrices $g^{(i)} = (\Omega_{jk}^{(i)})$, $i \in \mathbb{N}$, and the identity matrix (Example 6.34). The algebras $M_n(F)$ and $\mathrm{GM}_n(F)$ satisfy the same polynomial identities (Lemma 6.32).

The next theorem was established independently by E. Formanek and Yu. P. Razmyslov in the early 1970s.

Theorem 7.65 *For every $n \in \mathbb{N}$, there exists a central polynomial for $M_n(F)$.*

Proof Let $F(\Omega)$ be the field of rational functions in Ω (i.e., the field of quotients of $F[\Omega]$). Consider $\mathrm{GM}_n(F)$ as a subring of the $F(\Omega)$-algebra $M_n(F(\Omega))$. Denote by $\overline{\mathrm{GM}_n(F)}$ the $F(\Omega)$-subalgebra of $M_n(F(\Omega))$ generated by $\mathrm{GM}_n(F)$. This is, of course, just the linear span of $\mathrm{GM}_n(F)$. Corollary 7.46 states that the Capelli polynomial c_{n^2} is not an identity of $M_n(F)$, therefore it is not an identity of $\mathrm{GM}_n(F)$, and neither of $\overline{\mathrm{GM}_n(F)}$. Using Corollary 7.46 again it follows that $\overline{\mathrm{GM}_n(F)} = M_n(F(\Omega))$. This implies that $\mathrm{GM}_n(F)$ is a prime ring. Indeed, if $a, b \in \mathrm{GM}_n(F)$ satisfy $axb = 0$ for every $x \in \mathrm{GM}_n(F)$, then this obviously also holds for every $x \in \overline{\mathrm{GM}_n(F)} = M_n(F(\Omega))$, and hence $a = 0$ or $b = 0$ since $M_n(F(\Omega))$ is prime.

Thus, $\mathrm{GM}_n(F)$ is a prime PI-ring. The F-subalgebra of $\mathrm{GM}_n(F)$ generated by all generic matrices $g^{(i)}$ (but without the identity matrix) is clearly an ideal of $\mathrm{GM}_n(F)$. Theorem 7.56 implies that it contains a nonzero central element. Hence, the center of $\mathrm{GM}_n(F)$ contains elements different from scalar multiples of unity. The same therefore holds for the isomorphic algebra $F\langle X\rangle / I_n$. Let $f + I_n$ be an element of the center of $F\langle X\rangle / I_n$ which is not of the form $\alpha + I_n$, $\alpha \in F$. Note that $f - \alpha_0$, where α_0 is the constant term of f, is a central polynomial for $M_n(F)$.   □

With some extra effort one can prove Amitsur's theorem saying that $\mathrm{GM}_n(F)$ is not only prime, but a domain (see, e.g., [Row08, p. 467]). Consequently, $Q_Z(\mathrm{GM}_n(F))$ is also a domain; moreover, Theorem 7.58 thus implies that it is a finite dimensional central division algebra. It is called the **universal division algebra** of degree n.

Requiring the infinity of F in our proof of Theorem 7.65 is the price we had to pay for relying merely on the abstract theory. Constructive proofs, such as given by Formanek and Razmyslov, work regardless of the cardinality of F. (On the other hand, the case where F is finite can be reduced, by elementary means, to the case where F is infinite.)

As indicated in the last two paragraphs, there is much more to say. But every book must have an end.

Exercises

7.1. Let R be a ring such that its center Z is nonzero and all elements in $Z \setminus \{0\}$ are regular. Show that $Q_Z(S) \cong Q_Z(R)$ for every ring S such that $R \subseteq S \subseteq Q_Z(R)$.

7.2. Find an example of a ring R such that $R \subsetneq Q_Z(R) \subsetneq Q_{rc}(R)$.

7.3. Let $F \subseteq K$ be fields, and let T be a ring such that $F \subseteq T \subseteq K$ and K is the field of quotients of T. Note that $R = \begin{bmatrix} F & T \\ 0 & T \end{bmatrix}$ is a ring under the standard matrix operations. Prove that R satisfies the right Ore condition and describe its right classical ring of quotients. Show by an example that R does not necessarily satisfy the left Ore condition.

7.4. Show that every PI-algebra A satisfies a nonzero identity in two indeterminates. Hence derive that if A is a domain then it is a right (and left) Ore domain.

  Remark: The latter can be also deduced from Proposition 7.19, as well as from the structure theory (cf. Corollary 7.61).

7.5. Let I be a nonzero ideal of a prime ring R. Show that $Q_r(I) \cong Q_r(R)$.

7.6. Show that an automorphism of a nonzero prime ring R can be uniquely extended to an automorphism of $Q_r(R)$.

7.7. Let φ be an automorphism of a prime ring R, and let $0 \neq q \in Q_r(R)$ be such that $q\varphi(x) = xq$ for all $x \in R$. Show that q is invertible in $Q_r(R)$ (and so $\varphi(x) = q^{-1}xq$ for all $x \in R$).

7.8. Let $R \neq 0$ be a prime ring. Denote by $Q_s(R)$ the set of all $q \in Q_r(R)$ for which there exists a nonzero ideal J of R such that $Jq \subseteq R$. Show that $Q_s(R)$ is a subring of $Q_r(R)$ which contains R and whose center is the extended centroid C (thus, $R \subseteq R_C \subseteq Q_s(R) \subseteq Q_r(R)$). We call $Q_s(R)$ the **symmetric Martindale ring of quotients** of R. Note that q from the preceding exercise actually lies in $Q_s(R)$. Show that if R is a domain, then so is $Q_s(R)$.

7.9. Show that ξ is a zero-divisor in $Q_r(F\langle\xi, \eta\rangle)$. Thus, $Q_r(F\langle\xi, \eta\rangle)$ is not a domain and it properly contains $F\langle\xi, \eta\rangle$. Show, however, that $F\langle\xi, \eta\rangle$ is centrally closed.

7.10. Let R be a ring. Note that the set of all R-bimodule homomorphisms from R into R forms a ring under the standard operations. We denote it by W and call it the **centroid** of R. Show that:

   (a) If R is unital, then W is isomorphic to the center Z of R.
   (b) If $R^2 = R$, then W is commutative.
   (c) If R simple, then W is isomorphic to the extended centroid C of R.
   (d) If R is prime, then W can be embedded into C; show by an example that W is not necessarily isomorphic to C.

7.11. Let d and d' be derivations of a ring R. Check that $dd' - d'd$ is again a derivation, while there is no reason to believe that dd' is a derivation. In fact, show that if R is prime and $\operatorname{char}(R) \neq 2$, then dd' is a derivation only when $d = 0$ or $d' = 0$. The assumption on $\operatorname{char}(R)$ is necessary; show that if R is semiprime, then d^2 is a derivation if and only if $2d = 0$.

   Remark: The first of these two results was obtained by E. Posner in 1957, before the discovery of the extended centroid. However, the proof becomes shorter by making use of it.

7.12. Let R be a prime ring with extended centroid C. Suppose that a nonzero derivation d of R and $q \in Q_r(R) \setminus C$ satisfy $d(x)q = -qd(x)$ for all $x \in R$. Show that there exists $\lambda \in C$ such that $d(x) = \lambda[q, x]$ for all $x \in R$, and that $q^2 \in C$.

7.13. Let R be a ring. A biadditive map $D : R \times R \to R$ is called a **biderivation** if the maps $x \mapsto D(x, y)$ and $x \mapsto D(y, x)$ are derivations for every $y \in R$. Suppose that R is prime and noncommutative. Show that then there exists λ from the extended centroid C such that $D(x, y) = \lambda[x, y]$ for all $x, y \in R$. Hence derive that every additive map $f : R \to R$ satisfying $f(x)x = xf(x)$ for every $x \in R$ is of the form $f(x) = \lambda x + \mu(x)$ where $\lambda \in C$ and $\mu : R \to C$.

   Hint: Begin by computing $D(xy, zw)$ in two different ways.

7.14. Let R be a domain. Show that if R is a GPI-ring, then it is a PI-ring.

7.15. Let R be a simple unital ring. Prove that R is isomorphic to the ring of $n \times n$ matrices over a field if and only if there exists $0 \neq a \in R$ such that $axaya = ayaxa$ for all $x, y \in R$.

7.16. Let R be a prime ring, and let $a_i \in R$. Prove that $a_1 x a_2 x a_3 = 0$ for all $x \in R$ implies that at least one a_i is 0. Show by an example that $a_1 x a_2 x a_3 x a_4 = 0$ for all $x \in R$ does not, in general, imply that one of the a_i's is 0.

Hint: Theorem 7.51 suggests where to look for examples. However, they exist only under a strong restriction on $\operatorname{char}(R)$.

7.17. True or False:

(a) Every nonzero ideal of a prime PI-ring contains a regular element.
(b) Every nonzero ideal of a prime GPI-ring contains a regular element.
(c) Every left zero-divisor in a prime PI-ring is also a right zero-divisor.
(d) Every left zero-divisor in a prime GPI-ring is also a right zero-divisor.

7.18. True or False:

(a) If a nonzero ideal of a prime ring R is a PI-ring, then R is a PI-ring.
(b) If a nonzero left ideal of a prime ring R is a PI-ring, then R is a PI-ring.

7.19. Show that a prime PI-ring with finite center is isomorphic to the ring of $n \times n$ matrices over a finite field.

7.20. Let R be prime PI-ring, let $\widehat{Z}$ be the field of quotients of the center Z of R, and let $n \in \mathbb{N}$. Prove that the following statements are equivalent:

(i) $Q_Z(R)$ is a central simple algebra over $\widehat{Z}$ with $\dim_{\widehat{Z}} Q_Z(R) = n^2$.
(ii) R satisfies s_{2n}, but does not satisfy nonzero identities of degree less than $2n$.
(iii) There exists a field K such that R can be embedded in $M_n(K)$, and $M_n(K)$ satisfies the same multilinear identities as R (accordingly, R cannot be embedded in $M_{n-1}(H)$ for any commutative ring H).
(iv) The degree of algebraicity over $\widehat{Z}$ of every element in R is at most n, and there exist elements in R whose degree of algebraicity is n.

7.21. A polynomial $f(\xi_1, \ldots, \xi_m) \in F\langle \xi_1, \xi_2, \ldots \rangle$ is called a **weak identity** of $M_n(F)$ if $f(a_1, \ldots, a_m) = 0$ for all trace zero matrices $a_i \in M_n(F)$. Find a weak identity of $M_2(F)$ of degree 3.

Hint: Example 7.64.

7.22. Show that $[\xi_1, \xi_2]$ cannot be a central polynomial for a prime algebra.

Remark: This can be proved by applying the theory, but also directly, by elementary means.

7.23. Show that a central polynomial for $M_n(F)$ is an identity of $M_{n-1}(F)$.

References

[Bea99]   Beachy, J.A.: Introductory Lectures on Rings and Modules. London Mathematical Society Student Texts, vol. 47. Cambridge University Press, Cambridge (1999)

[BMM96]   Beidar, K.I., Martindale 3rd, W.S., Mikhalev, A.V.: Rings with Generalized Identities. Monographs and Textbooks in Pure and Applied Mathematics, vol. 196. Marcel Dekker Inc, New York (1996)

[FD93]   Farb, B., Dennis, A.V.: Noncommutative Algebra. Graduate Texts in Mathematics, vol. 144. Springer, Berlin (1993)

[Her68]   Herstein, I.N.: Noncommutative Rings. Carus Monographs in Mathematics, vol. 15. The Mathematical Association of America, Eau Claire (1968)

[Hun74]   Hungerford, T.W.: Algebra. Graduate Texts in Mathematics, vol. 73. Springer, Berlin (1974)

[Lam01]   Lam, T.Y.: A First Course in Noncommutative Rings. Graduate Texts in Mathematics, vol. 131, 2nd edn. Springer, New York (2001)

[Lam95]   Lam, T.Y.: Exercises in Classical Ring Theory. Problem Books in Mathematics. Springer, New York (1995)

[Lam99]   Lam, T.Y.: Lectures on Modules and Rings. Graduate Texts in Mathematics, vol. 189. Springer, New York (1999)

[MR01]   McConnell, J.C., Robson, J.C.: Noncommutative Noetherian Rings. Revised Edition. Graduate Studies in Mathematics, vol. 30. American Mathematical Society, Providence (2001)

[McC04]   McCrimmon, K.: A Taste of Jordan Algebras. Universitext. Springer, New York (2004)

[Pie82]   Pierce, R.S.: Associative Algebras. Graduate Texts in Mathematics, vol. 189. Springer, New York (1982)

[Row91]   Rowen, L.H.: Ring Theory, Student edn. Academic Press, San Diego (1991)

[Row08]   Rowen, L.H.: Graduate Algebra: Noncommutative View. Graduate Studies in Mathematics, vol. 91. American Mathematical Society, Providence (2008)

[ZSSS82] Zhevlakov, K.A., Slinko, A.M., Shestakov, I.P., Shirshov, A.I.: Rings that are Nearly Associative. Pure and Applied Mathematics, vol. 104. Academic Press, New York (1982)

© Springer International Publishing Switzerland 2014   193
M. Brešar, *Introduction to Noncommutative Algebra*, Universitext,
DOI 10.1007/978-3-319-08693-4

Index

A

Abelian group, xxii
Additive group, xxii
Additive map, xxiii
Additive subgroup, xxvi
Algebra, xxxiii
Algebra homomorphism, xxxiv
Algebraic algebra, 5
Algebraic algebra of bounded degree, 160
Algebraic closure, xxxvii
Algebraic element, xxxvi, 5
Algebraically closed field, xxxvii
Alternating polynomial, 143
Alternative algebra, 5
Amitsur, S. A., 128, 157, 160, 181, 189
Amitsur-Levitzki Theorem, 143, 153, 157, 161, 187
Annihilator (left, right) of a subset of a ring, 76
Annihilator of a module, 56
Antiautomorphism, 23
Antiisomorphism, 63
Artin, E., 71, 115
Artin-Whaples Theorem, 115, 134
Artinian module, 72, 76, 77
Artinian ring, 72, 73, 77
Ascending chain condition, 72
Augmentation ideal, 36, 51
Automorphism, xxiii, xxvi, 57
Axiom of choice, xxi, 72

B

Base field, xxxvi
Basis, xxxii, 59
Bergman, G., 159
Biadditive map, xxiii

B (continued)

Biderivation, 190
Bilinear map, xxxii
Bimodule, 54
Boolean ring, 131, 147, 184
Brauer group, 103
Brauer, R., 103
Burnside's theorem, 115, 134
Burnside, W., 115

C

Capelli polynomial, 144, 179, 188
Cartesian product, xx
Cayley's theorem, xxiii, 33
Cayley-Hamilton Theorem, 154, 161, 188
Center of a group, xxiii
Center of a ring, xxv
Center of an algebra, xxxiv
Central algebra, 9
Central closure, 175, 176, 181, 183, 185
Central element, xxv
Central idempotent, 40
Central polynomial, 188, 191
Central simple algebra, 10, 12, 13, 18, 22, 44, 46, 48, 49, 51, 91, 92, 94–98, 100–102, 104–106, 151, 178, 185–187, 191
Centralizer, 92, 94, 99, 100, 106
Centrally closed, 176, 179, 190
Centroid, 190
Chain, xx
Characteristic of a field, xxvi
Characteristic of a prime ring, 29
Characteristic of a ring, xxvi
Characteristic of an algebra, xxxiv
Characteristic polynomial, 154
Class formula, 17

© Springer International Publishing Switzerland 2014
M. Brešar, *Introduction to Noncommutative Algebra*, Universitext,
DOI 10.1007/978-3-319-08693-4